CLIMATE CHANGE POLITICS IN EUROPE

Climate Change Politics in Europe

Germany and the International Relations of the Environment

LYN JAGGARD

Tauris Academic Studies

LONDON • NEW YORK

Published in 2007 by Tauris Academic Studies, an imprint of I.B.Tauris & Co Ltd
6 Salem Road, London, W2 4BU
175 Fifthe Avenue, New York NY 10010
www.ibtauris.com

In the United States of America and Canada distributed by Palgrave Macmillan
a division of St. Martin's Press, 175 Fifth Avenue, New York NY 10010

ISBN: 978 1 84511 409 1

A full CIP record for this book is available from the British Library
A full CIP record is available from the Library of Congress

Library of Congress Catalog Card Number: available

Printed and bound in India by Replika Press Pvt. Ltd
From camera-ready copy edited and supplied by the author

CONTENTS

NOTE ON SPELLING

In words such as realise and globalisation 's' has been used except where a direct quote has been reproduced and 'z' has been used in the original text. The same applies to cooperation and co-operation.

ABBREVIATIONS

AA	*Auswärtiges Amt* - Foreign Office
AIJ	Activities Implemented Jointly
AOSIS	Alliance of Small Island States
BDI	*Bundesverband der Deutschen Industrie* - Federal Association for German Industry
BMBF	*Bundesministerium für Bildung und Forschung* - Federal Ministry for Education and Research
BMF	*Bundesministerium der Finanzen* - Federal Ministry for Finance
BMU	*Bundesministerium für Umwelt, Naturschutz und Reaktorsicherheit* - Federal Ministry for Environment, Nature Conservation and Nuclear Safety
BMWA	*Bundesministerium für Wirtschaft und Arbeit* - Federal Ministry for Economics and Labour
BMZ	*Bundesministerium für wirtschaftliche Zusammenarbeit und Entwicklung* - Federal Ministry for Economic Cooperation and Development
BUND	*Bund für Umwelt und Naturschutz Deutschland - Freunde der Erde* - German Federation for Environment and Nature Protection - Friends of the Earth Germany
CDM	Clean Development Mechanism
CDU	*Christlich Demokratische Union* - Christian Democratic Union
CFC	Chlorofluorocarbons
CH_4	Methane
CO_2	Carbon dioxide
COP	Conference of the Parties
CSD	Commission for Sustainable Development
CSU	*Christlich-Soziale Union* - Christian-Social Union
DDT	Dichloro-diphenyl-trichloro-ethane

DEFRA	Department of Environment, Food and Rural Affairs (UK)
EC	European Community
ECCP	European Climate Change Programme
EEC	European Economic Community
ENB	Earth Negotiations Bulletin
EU	European Union
FDP	*Freie Demokratische Partei* - Free Democratic Party
FIELD	Foundation for International Environmental Law and Development
FSM	Federated States of Micronesia
GEF	Global Environment Facility
GHG	Greenhouse gas
GTZ	*Deutsche Gesellschaft für Technische Zusammenarbeit* - German Society for Technical Cooperation
H-CFC	Hydrogenous chlorofluorocarbons
KfW	*Kreditanstalt für Wiederaufbau* - Credit institution for reconstruction
INC	Intergovernmental Negotiating Committee for the Framework Convention on Climate Change
IPCC	Intergovernmental Panel on Climate Change
JI	Joint Initiative
JREC	Johannesburg Renewable Energy Coalition
LULUCF	Land Use, Land-use Change and Forestry
N_2O	Dinitrogen oxide
NGO	Non-governmental organisation
OECD	Organisation for Economic Cooperation and Development
ODA	Overseas development assistance
OPEC	Organisation of Petroleum Exporting Countries
PICCAP	Pacific Island Climate Change Assistance Programme
PIREP	Pacific Islands Renewable Energy Project
PIK	*Potsdam Institut für Klimafolgenforschung* - Potsdam Institute for Climate Impact Research
POPs	Persistent organic pollutants
RMI	Republic of the Marshall Islands
SBI	Subsidiary Body for Implementation (UNFCCC)

SBSTA	Subsidiary Body for Scientific and Technological Advice (UNFCCC)
SEF	*Stiftung Entwicklung und Frieden* – Development and Peace Foundation
SF_6	Sulphur hexafluoride
SIDS	Small island developing States
SPD	*Sozialdemokratische Partei Deutschlands* - Social Democratic Party of Germany
UN	United Nations
UNCED	United Nations Conference on Environment and Development
UNDP	United Nations Development Programme
UNEP	United Nations Environment Programme
UNFCCC	United Nations Framework Convention on Climate Change
UNGA	United Nations General Assembly
WBGU	*Wissenschaftlicher Beirat der Bundesregierung Globale Umweltveränderungen* - German Advisory Council on Global Change
WEO	World Environmental Organisation
WMO	World Meteorological Organisation
WTO	World Trade Organisation
WWF	World Wide Fund for Nature

1

INTRODUCTION

The purpose of this study is to analyse the role that Germany plays within the international relations of climate change. The way in which Germany conducts these relations and any influence that it has will be examined.

Conceptual framework and theory

Climate change and its anthropogenic causes are transnational in nature; as such neither the causes nor the consequences can be managed by individual nation-states acting alone. It could be argued that climate change is the ultimate example of globalisation; it is a phenomenon that will affect all parts of the world, albeit to varying extents and in varying ways. Whilst it is essential that individual countries implement a variety of policies in order to manage the effects of climate change and to try to minimise future adverse effects, the transboundary nature of this phenomenon necessitates international political cooperation so that this threat to global environmental security can be managed. The United Nations Framework Convention on Climate Change (UNFCCC) facilitates such international cooperation and provides a framework through which nation-states act domestically to reduce greenhouse gas emissions (GHGs).[1] The UNFCCC was opened for signature at the United Nations Conference on Environment and Development (UNCED), which was held in Rio de Janeiro in June 1992. The need for sustainable development was also addressed at UNCED. Pursuing sustainable development and addressing climate change are interlinked. What is termed the 'developed world' is coterminous with that part of the world that has industrialised, however, emissions from industrial processes are the major cause of climate change. If sustainable development is to be achieved, the problems and causes of climate change must be addressed. The World Summit on Sustainable Development (WSSD) was held in 2002, it was the ten year follow up to the Rio conference and a major part of the continuing UNCED process. It is for this reason that the WSSD has been chosen as a case study. The eighth Conference of the Parties

(COP8) to the UNFCCC was held shortly after the WSSD, it is therefore appropriate that events at COP8 are also examined. Both conferences were major arenas for continuing climate change international relations; it will be seen that Germany was an active participant at these conferences.

Whilst climate change requires some form of global governance, the world does not have a global government; the accepted form of global order consists of the international relations of nation-states. Germany is an industrialised country and as such it has contributed to climate change. It is however, a country that is currently trying to actively contribute to both the climate change governmental process and to improving the environmental footprints that it makes.

As is the case in all matters, the way climate change politics are conducted is influenced by the way in which the problem and its possible solutions are perceived. The influences that values, ideas and concerns have had in the evolution of Germany's climate change related policies will be examined; it will be shown that these have in turn impacted on Germany's aims in international relations. These aims and the avenues that Germany takes to try to achieve these aims will be examined and compared with the outcomes of climate change related international relations; analysis will determine where and to what extent Germany has managed to achieve influence. Textual analysis of documentation, along with information gleaned from participants of this process will enable this analysis. Correlations of opinions do not necessarily prove a causal link and the analysis will separate issues where influence can be definitely shown from those where German influence is not absolutely identifiable.

Climate change, its anthropogenic causes and its transnational but varying effects clearly involve injustices. The benefits of industrialisation have accrued to those countries and people that are least vulnerable to climate change, however they are the most able to afford to implement adaptation policies. Conversely those areas of the world that are affected most detrimentally by climate change are those that tend not to have received the benefits of industrialisation. The small low-lying islands of the world are amongst the first and worst to be affected. Some Pacific Islands are already being detrimentally affected; they have contributed very little, if at all to the environmental degradation that has caused their plight. Inter-societal injustice exists. The existence of future generations will be fundamentally affected by the actions that are taken now. Clearly questions of intergenerational justice are relevant. As already stated, climate change is a phenomenon that is inherently global and which therefore requires international negotiations and institutions to provide governance. To enable agreement on what action needs to be taken requires a great deal of discussion between States and a host of scientists, politicians and citizens from around the world. Agreements at the international level need to be consensual; Germany's domestic politics is also noted for its consensual nature. It will be seen that a variety of actors, such as research institutions and non-governmental organisations, are involved within climate change political processes in Germany. Extensive discussion is often required to reach a consensus. These discursive imperatives have led me to employ Jürgen Habermas' work on discourse ethics in my analysis. In essence Habermas calls for participatory justice and the extent to which this is achieved both in Germany and the international arena will be assessed. The acceptability of the terms and inclusiveness of discussions and hence the perceived justice of the results are important, as policies are more likely to succeed if everyone agrees with them, i.e. if they are reflective of the ideas of the participants, who have through discussion, arrived at a consensual decision. As detailed above, other forms of justice considerations are also relevant and they are evidenced by the inclusion of the polluter-pays principle (retributive justice) and the precautionary principle (to try to ensure intergenerational justice), in both German environmental policy-making and in international agreements.

The works of Jürgen Habermas and to a lesser extent those of Henry Shue will be considered. Habermas (1996) talks of a system of rights when discussing a reconstructive approach to law. This seems apt as essentially the climate change regime is looking toward an international law to govern and mitigate the effects of climate change. Henry Shue (1980) identifies some basic rights, which include: physical security; subsistence (minimal economic security); unpolluted air and unpolluted water. Shue also holds that for every basic right there are three types of duty, these are: duties to *avoid* depriving; duties to *protect* from deprivation; and duties to *aid* the deprived. Habermas' theory emphasises a discursive approach through which agreement can be reached to formulate a system of rights and duties; this will be used to help analyse climate change political processes in Germany and the international arena.

It is both important and appropriate to use Habermas' work on discourse ethics in order to understand Germany's international relations of climate change as the reasons given above for choosing this approach illustrate. It is also the case that as Germany's domestic politics are consensual and discursively inclusive there is an incentive for Germany to pursue ethical discourse at the international level as German negotiators may have a competitive advantage due to communicative abilities and discursive routines honed within domestic politics.

International institutions are formed to enable States to cooperate with one another in order to achieve positive outcomes for all concerned. Cooperation requires discursive processes to be entered into; whilst these processes need not necessarily comply with discourse ethics, the holding of discussions is a basic requirement and could be the first step toward such an eventuality and thus Habermasian analysis is relevant. The idea of pursuing positive outcomes for all is opposed to the traditional realist (in the international relations sense of the word) viewpoint that States are interested only in relative gains, i.e. that their own power vis-à-vis other States is not compromised. Neo-liberal institutionalist theory is relevant to this study in so far as the premise behind the UNCED and UNFCCC processes is that by working together all will benefit, this is imperative in the case of climate change politics as it is a domain in which States must cooperate if policies are to prove effective. Institutions are formed in order to manage complex interdependence. Keohane and Nye (1989) describe complex interdependence as a situation in which interconnections that create interdependencies abound between many societies. The dependences may not be equal but they exist. This leads to the diminishing validity of the traditional idea of the threat of force being used to achieve ones aims. Neo-liberal institutionalism[2] recognises the importance of input from States, and from other actors such as the scientific community, epistemic communities and non-governmental organisations.

Realist theory[3] holds that the key actors in international relations are still States, other actors may have input to a variety of institutions but that it is still States that have the authority to make decisions and take action. Neo-liberal institutionalism agrees that States are still very important players in the world order but sees the importance of other actors as being more instrumental than do realists. The emphasis in institutionalism is on consensual politics, as opposed to realism, which focuses on power politics. International climate change agreements are made by States and some elements of realist theory will be seen to be relevant, especially with regard to the action of some participants. However, consensus needs to be found in order to reach international agreement and this coincides with neo-liberal institutionalist theory. The need for consensus is also recognised by Habermas who calls for this to be reached through multi-participative ethical discourse.

This study focuses on climate change international relations; however, justice considerations are inherent in this subject matter and they will also be considered. Habermas' discourse ethics will be used in the analysis of international relations, and some elements of neo-liberal institutionalism and realism will also be assessed.

Climate change and the various related interconnections between a variety of States, institutions and peoples, clearly occur; some of these connections are observable, some are extremely difficult if not impossible to observe. Relationships can be interpreted in various ways. The way in which negotiators view such relationships will have an affect on the attitudes, aims and strategies that they bring to the negotiating table. As climate change and associated international relations exist, the ontological position that I take is foundationalist. Hughes and Sharrock explain foundationalism as

> the view that true knowledge must rest upon a set of firm, unquestionable set of indisputable truths from which our beliefs may be logically deduced, so retaining the truth value of the foundational premises from which they follow, and in terms of which our methods of forming further ideas about the world and investigating it can be licensed. (Hughes and Sharrock, 1997:4-5)

It should be pointed out that Habermas has chosen not to recognise an ontological position. Broadly defined Habermas works within the critical theory paradigm.[4] Critical theory ontologically and epistemologically aligns with what is termed critical realism, which

> straddles both the positivist and interpretivist paradigms, sharing a foundationalist ontology with positivism and by allowing for interpretation in research. (Grix, 2004:86)

Epistemologically, if one takes positivism, realism (of the social science variety) and relativism to exist along a continuum then the position I take is realist, but a realism that is fairly close to the relativist end of the continuum.[5] I believe that how phenomena are interpreted is important and does impact on events, however, I also believe that there is a reality beneath the surface even if it cannot be fully explained.

The ontological and epistemological positions that I identify are consistent with the views of Habermas. Although constructivism is an important element in Habermas' critical theory, it is also the case that as argued by How when discussing Marcuse and Habermas

> for Critical Theory, explanation has to be grounded in something real, something that exists in some degree independent of our descriptions of it, (How, 2003:155)

It is perhaps worth noting that the German political system and the way in which Germany conducted relations with the international community had to be rethought and re-established after the Second World War. A largely consensual political system has resulted and international relations are conducted in a discursive manner that aims to lead to consensus. Habermas was born in Germany in 1929; his adult life and his political opinions would at least in part have been influenced by

Germany's political evolution and transformation. Habermas' theoretical perspective will be expanded upon in the next chapter. However, it is appropriate at this stage to outline questions arrived at from a Habermasian outlook that can be asked of the German political system and the German approach to international climate change relations. Is multi-participation in policy-making encouraged and implemented? Are discussions inclusive of all interested and affected parties? Are decisions arrived at through rational decision-making i.e. are decisions the result of the unforced force of the better argument? If all interested parties agree on principles that are justifiable and universalisable, are they institutionalised, thus creating rights and responsibilities?

Habermas talks of an ideal speech situation in which all interested and affected parties participate in discourse. The contributions of all parties should be equally respected in terms of all contributions being equally considered through reasoned and rational discussion. All parties should have linguistic and communicative abilities. It is recognised that this is an ideal and that in reality strategic interests do enter into the equation. As Stokke points out

> Habermas does not deny that humans are able to act strategically, i.e. in a goal-seeking manner towards objectified others, in order to realize a desired condition; the point is rather that their actions *cannot be reduced* to this purpose. There is something more going on which the actor himself cannot avoid or fail to notice: action expresses inner feelings and sentiments; it relates to social norms, either approvingly or rejectingly; and it communicates an understanding of the world. (Stokke, 1998:135)

This discussion indicates that from a Habermasian point of view one might expect: German climate change related policies to be the result of rational decision-making arrived at through discourse that includes a multiplicity of inputs and considerations; and for Germany's international relations of climate change to encourage and facilitate participative and just discursive processes.

Research sources

As Habermas' discourse ethics is used in analysis it is appropriate to examine a wide variety of sources in order to determine the participatory inclusiveness of the German political system and the extent to which multi-participation is encouraged in and through Germany's international relations of climate change.

Research for this study has involved the reading and analysis of: theoretical texts; literature examining the evolution of environmental politics in Germany, the European Union (EU), and at the international level, especially those relating to climate change; and documentation relating to international agreements. Discursive processes can exist between a variety of organisations and communities. To help establish the extent of participatory inclusiveness in the decision-making process of the official German position on the international relations of climate change, the positions of selected research institutions, and non-governmental organisations will be analysed and connections between these and the government position assessed. Interactions between industry and the German government and also the opportunity for the general public to contribute opinions on and to the government position will also be examined. The governmental process will also be considered. Reports from, and information gleaned from the websites of institutions, organisations and the German government will be studied, along with information gained through interviews. The events up to, during and following the WSSD held in Johannesburg in August/September 2002, and COP8 held in New Delhi in October/November 2002 were monitored. The conferences websites were monitored, as were on-line Earth Negotiations Bulletin reports and numerous on-line newspaper reports. Examination of a variety of literature and documentation has enabled assessment of Germany's aims and strategies with regard to the WSSD and COP8. Evaluation of the outputs from these conferences has helped determine the extent to which Germany influenced outcomes. To enable further understanding of the way in which Germany conducts its international relations and to determine the discursive inclusiveness it promotes, international relations that Germany conducts away from international gatherings will be examined. Semi-structured interviews are the main source that have enabled this analysis, however, written information from a variety of sources including the: Global Environment Facility; International Conference for Renewable Energy; UNFCCC; Wuppertal Institute for Climate, Environment and Energy; and the Pacific Islands Forum was also accessed.

Fieldwork in Germany was conducted between January and May 2003. Berlin was the base for fieldwork and where the vast majority of research was undertaken. However, research was also conducted in Bonn and Potsdam.

Fieldwork included:

➢ Time spent at the *Wissenschaftzentrum Berlin für Sozialforschung* (WZB) (Social Science Research Centre Berlin). Professor Simonis, Research Professor for environmental policy at the WZB, provided me with academic support during my stay, along with the practical support of a place to work. A great deal of relevant literature was researched in the WZB library.

➢ Time spent at the *Wissenschaftlicher Beirat der Bundesregierung Globale Umweltveränderungen* (WBGU) (German Advisory Council on Global Change to the Federal Government). Staff at the WBGU provided some valuable insights into the German political and climate change processes. The WBGU also has a small, but dedicated library, which also proved of great use.
➢ The UNFCCC secretariat was visited, and time was spent in the library researching documentation. A number of interviews were also conducted at the secretariat.
➢ During the course of my research in excess of thirty semi-structured interviews were conducted, the vast majority of which were in Germany. Interviewees included representatives of: government departments; research institutions; non-governmental organisations; and universities. E-mail correspondence was also conducted with representatives of various organisations.

A research trip to Brussels was made in February 2004, where representatives from the European Commission's Directorate General for the Environment and Directorate General for Energy and Transport, were interviewed. In Britain, interviews were also conducted with a representative of the UK's Department of Environment, Food and Rural Affairs and a representative of the Foundation for International Environmental Law and Development. These interviews enabled views from a 'non-German' perspective to be considered, thus enabling me to achieve triangulation.

Chapter outline

Chapter two will provide a theoretical overview that underpins analysis of the empirical body of this study. Theoretical analysis will of course be interwoven throughout the text.

Chapter three will include the evolution of environmental policies in Germany since the 1970s. Environmental politics have become an integral part of both national and global political arenas in recent years. In many instances environmental issues are transnational; hence the national and international are often interlinked in environmental politics. Bearing this in mind, the text will concentrate on the development of environmental awareness and politics in Germany since the 1970s when issues of environmental degradation really came to the fore. However, a brief background of environmental issues and attitudes prior to the 1970s will be included, as will some international links. It will, for example, be seen that there are connections between domestic policies and the 1972 UN Conference on the Human Environment held in Stockholm. This analysis will be largely in the form of a literature review. This will be followed by examination of the evolution of global climate change politics, i.e. the 1992 UNCED and the UNFCCC. The continuing UNCED process will be briefly outlined. It will be noted that reflexivity between the various levels does occur.

Chapter four will focus on the formulation of Germany's aims for the WSSD. COP8 was held within two months of the WSSD and therefore the events leading up to and at the WSSD are relevant to aims for COP8. This chapter will analyse various probable inputs to the government position. These include inputs from, research institutions, non-governmental organisations, industry, and various government departments. The extent to which domestic policies impact on international preferences and the idea of leading by example will be investigated.

This study concentrates on Germany's international relations of climate change. However, Germany often works with and through the EU, thus chapter five will provide a synopsis of the evolution of climate change related politics in the EU. At conferences such as the WSSD and COP8 the EU presents a united position, therefore, the development of the EU stance will be studied. Germany's role in the formulation of the EU position will also be examined.

Events at, and outcomes of the WSSD and COP8 will be analysed in chapter six. Germany's aims will be compared with the outcomes of these conferences and areas of influence that Germany achieved will be assessed. Implications for Germany of the outcomes of these conferences will also be considered.

Chapter seven will examine Germany's international relations of climate change that have been conducted in addition to the WSSD and COP8. Germany's relations with and through the UNFCCC process and the Global Environmental Facility will be investigated, as will a variety of other multilateral relations. Bilateral relations will also be considered.

In order to assess the success or otherwise of Germany's policies in the international relations of climate change, analysis of any positive (and negative) affects on Pacific Island countries will be undertaken. As previously mentioned, low-lying Pacific Island countries are particularly vulnerable to climate change and it is therefore, appropriate to look at whether or not Germany's international relations prove to be of any benefit to this group of States.

Germany's role in the international relations of climate change will be briefly summarised in the conclusion. Also included will be an assessment of how the preceding material provides corroboration for the assertion that it is appropriate to use Habermas' discourse ethics in evaluation.

2

THEORY OVERVIEW

Justice considerations were identified in the introduction as being relevant to climate change international relations and to be used as a tool of analysis. Clarification of the definition of the term justice is required. A dictionary definition of justice is:

> 1. the quality of being just or fair;
> 2. the principle of fair treatment or conduct;
> 3. the administration of the law;
> 4. a judge or magistrate. (Harber & Payton, 1989:562)[1]

The term 'just' includes the definitions: fair; even-handed; impartial; honest; and morally proper.

In relation to this study the first two definitions are those that will be concentrated upon. Administration of the law may be touched upon in relation to the formalising of agreements into international law such as the legally binding Kyoto Protocol; however this element of justice will not be analysed. It will be the process of coming to agreement and enshrining such in international law that will be the focus of justice analysis. Clearly the fourth definition does not apply to this research.

In short the definition of justice that is the starting point for this study is: *the qualities and principles of fairness; even-handedness i.e. equality; and moral decency.* Moral justice is being used as an analytic tool not legal justice per se; though how to get moral justice established as legal justice is an element of the research. Problems arise in the fact that terms such as moral decency and morally proper are open to contestation. Events and outcomes that are seen as fair and just by some people can be seen completely differently by another group of people. The international relations of climate change involve diplomacy and discussions in attempt to gain agreement and hence progress towards mitigating and adapting to climate change. Hence the appropriateness of discourse based theories with the aim of arriving at agreed rights and responsibilities. Cavalier *et al* discuss Habermas's views on discourse ethics during which they state that

> Habermas's central point [is]: communicative action defines a rationality capable, through discourse, of arriving at universal norms. (Cavalier *et al,* 2002)

Habermas holds that during discourse

> Every subject with the competence to speak and act is allowed to take part in a discourse.
> Everyone is allowed to introduce any assertion whatever into the discourse.
> Everyone is allowed to express his attitudes, desires, and needs.
> No speaker may be prevented, by internal or external coercion, from exercising his rights … (Ibid)

This idea of everyone being able to participate in discussions in order to find a consensus on the best course of action is a principle that can be seen, at least to some degree, to be in action within Germany. Whilst not everyone has, nor could possibly have, equal access to discursive processes that result in decision-making; the fact that Germany has a consensually orientated political system does mean that a wide variety of opinions are considered.[2]

As mentioned above, rationality is a key factor behind the belief that complex issues can be discussed and consensus found. The belief that human rationality can determine future paths of development can be traced back to the Enlightenment. Immanuel Kant philosophised on the subject of reason, Onora O'Neill explains what she understands to be Kant's thoughts on the meaning of reason.

> 1. To think for oneself;
> 2. To think from the standpoint of everyone else; and
> 3. Always to think consistently. (Rossi, 1998:72)

Kant proposes that by reasoning, people can act in ways that would be deemed proper for all people to act in that same situation, i.e. such actions are justifiable. By thinking in this way principles can be agreed upon that are justifiable and universalisable. Kant calls this the Categorical Imperative. According to McCarthy

> Habermas's discourse model represents a procedural reinterpretation of Kant's categorical imperative: rather than

> ascribing as valid to all others any maxim that I can will to be a universal law, I must submit my maxim to all others for purposes of discursively testing its claim to universality. The emphasis shifts from what each can will without contradiction to be a general law, to what all can will to be a universal norm. (McCarthy, 1984:326)

Thus when all can achieve consensus, a universal norm is created. If all have agreed on an action it will be perceived as being a just decision, therefore, all will be willing to abide by such a decision and any following actions. In other words, participatory justice will have been fulfilled. In some instances decisions arrived at consensually are institutionalised in law, i.e. in rights and responsibilities. Once rational discussion has brought about consensus on universal norms, such consensus can be enshrined in rights. To quote Shue

> A moral right provides the rational basis for a justified demand that the actual enjoyment of a substance be socially guaranteed against threats. (Shue, 1980:13)

Habermas discusses rights in which he states that individual rights are important in understanding modern law. He states that

> rights ("subjective rights" in German) fix the limits within which a subject is entitled to freely exercise her will. More specifically, they define the same liberties for all individuals or legal persons understood as bearers of rights. (Habermas, 1996:82)

Habermas links the idea of individual rights, society and discourse in the following quote

> the procedure of democratic legislation must confront participants with the normative expectation of an orientation to the common good, because this procedure can draw its legitimating force only from a process in which citizens *reach an understanding* about the rules for them living together. In modern societies as well, the law can fulfil the function of stabilizing behavioral expectations only if it preserves an internal connection with the socially integrating force of communicative action. (Ibid, pp83-84)

Habermas talks of democratic legislation and hence this extract clearly pertains to a 'democratic community'; most people would probably think first of a nation-state. Clearly people within any community need to be able to get along in a manner that is acceptable to the majority of people within that community. If one takes into account the global nature of climate change and the required international negotiations, it can be seen that a much larger global community is in evidence. Thus the above quote can be seen to be relevant to this transnational community. International climate change negotiations aim to '*reach an understanding* about the rules for them living together' (Ibid) the aim is to enshrine in international law 'behavioral expectations … with the socially integrating force of communicative action.' (Ibid)

Basic rights that are identified by Habermas are

1. Basic rights that result from the politically autonomous elaboration of the *right to the greatest possible measure of equal individual liberties.*

These rights require the following as necessary corollaries:

2. Basic rights that result from the politically autonomous elaboration of the *status of a member* in a voluntary association of consociates under law.
3. Basic rights that result immediately from the *actionability* of rights and from the politically autonomous elaboration of individual *legal protection.*

… only with the next step do legal subjects also become *authors* of their legal order, to be exact, through the following:

4. Basic rights to equal opportunities to participate in processes of opinion-and will-formation in which citizens exercise their *political autonomy.*

…the rights listed so far *imply* the following:

5. Basic rights to the provision of living conditions that are socially, technologically, and ecologically safeguarded, in so far as the current circumstances make this necessary if citizens are to have equal opportunities to utilize the civil rights listed in (1) through (4). (Ibid, pp122-123)

If all people have certain rights, then they also have responsibility to ensure that all people have those rights fulfilled. As mentioned in the introduction, Shue (1980) argues that for every basic right there are three correlative duties: to *avoid* depriving; to *protect* from deprivation; and to *aid* the deprived.

The implementation of these responsibilities or duties is necessary to fulfil recognised universal rights, however, it is not always the case that such responsibilities are met. Examples of this can be found in the United Nations Universal Declaration of Human Rights, which was adopted on

10 December 1948 by the UN General Assembly. Only a few of these rights shall be reproduced here in order to demonstrate that universally agreed rights are not always universally fulfilled.[3]

> Article 4. No one shall be held in slavery or servitude; slavery and the slave trade shall be prohibited in all their forms.
> Article 5. No one shall be subjected to torture or to cruel inhuman or degrading treatment or punishment.
> Article 25. (1) Everyone has the right to a standard of living adequate for the health and well-being of themselves and their family, including food, clothing, housing and medical care and necessary social services, and the right to security in the event of unemployment, sickness, disability, widowhood, old age or other lack of livelihood in circumstances beyond their control.
> (www.un.org/Overview/rights.html June 2002)

Article 25 (1) is clearly not being completely fulfilled even in wealthy industrialised countries. Climate change is very likely to detrimentally effect the provision of this right to millions of people worldwide. It is vitally important that through discussion the international climate change negotiations can arrive at universally agreed courses of action and rights, that can be translated into international law and that systems are put in place to ensure compliance of these rights and/or laws.

Although Habermas details the process through which universal norms can be determined, it is not the case that such norms are always achievable, as not all interests are generalisable. The main point is that ethical discourse[4] be engaged in and that where a universal norm is not achievable, a consensually agreed upon compromise should be reached. In this way democratic will formation can be achieved. It should be noted that compromise requires changing actions, whilst consensus can imply changing cognitions, interest patterns and options for actions.

The idea of democratic will formation can be applied to various different situations and is appropriate to be applied to the formation of Germany's aims in the international relations of climate change and to the way in which Germany's conducts its climate change international relations. Discussing Habermas, McCarthy states that

> democracy, as a principle of political order, does not single out a priori one specific type of organization … The point is, rather, to find in each set of concrete circumstances institutional arrangements that justify the presumption that basic political decisions would meet with the agreement of all those affected by

> them if they were able to participate without restriction in discursive will formation. (McCarthy, 1984: 332)

Multi-participatory discursive processes that are used to determine democratic will formation can be implemented at multiple and overlapping levels. The following chapters will examine interactions between German sub-national groups and the government in the formation of Germany's policy with regard to the international relations of climate change. Germany's relations with the European Union, whilst not the focus of this study will also be examined and Germany's relations with the wider international community will be analysed.

As already stated the discursive nature of climate change international relations and the need to find common agreement for the future direction of global climate change related policies means that discourse ethics as propounded by Habermas is appropriate to help analysis. If agreements are reached through discourse where participatory justice has been implemented, agreements are more likely to be acceptable to all parties and thus the prospects for successful implementation will be greater.

Neo-liberal institutionalism and realism are conventional approaches that are often used in international relations analysis. Elements of both of these theories would appear to be appropriate to certain aspects of climate change international relations. Neo-liberal institutionalism may appear to best describe the actions of many participants and it may be that global negotiations are progressively turning toward cooperation for mutual advantage; however although neo-liberal institutionalism acknowledges the multi-participative nature of international relations and looks towards all benefiting, it does not adequately address the need for equitable participation and thus the achievement of equitable and universally acceptable outcomes. It would also appear that the USA's actions within the climate change regime are a prime example of a State wanting to exercise its power to the greatest possible advantage for itself. The attitude of the USA's government in many way fits into the classical realist framework of an hegemonic power pursuing self-interest with very little consideration being given to the opinions of 'lesser' powers. Justice considerations may well be made by all parties, although the perceptions of justice appear to have a different focus. Participatory justice is the core ideal through which to reach consensus through ethical discourse; the implications of this are that through equitable and just processes, equitable and just outcomes are likely to occur. Neo-liberal institutionalism is concerned with multi-participation and determining the best way to implement win-win scenarios, thus resulting in relatively equitable outcomes and justice for all being considered to some extent. States that operate in a realist manner are more interested in justice for themselves, i.e. the USA is concerned that international climate change agreements will detrimentally affect their economy and lifestyles. Whilst neo-liberal institutionalism and realism may explain some elements of climate change international relations, it can be argued that the use of Habermas' discourse ethics has greater relevance and utility. It can: lead to greater understanding of how interests are formed and legitimised; facilitate analysis of actions taken; and provide a framework through which just and implementable decisions can be reached.

Conventional analytical approaches tend to view power in terms of military and/or economic might. Although these attributes do confer the ability to exercise power, it is simplistic to look only to these features in order to explain events. Many influences can contribute to outcomes, albeit to different extents. To be able to influence matters is in itself a form of power. The phenomenon of climate change was initially introduced to the world by scientists, media attention created public and political awareness; to get a subject onto the agenda for discussion is itself a form of power being exercised.[5] Everyone who has access to information and discursive inputs can play a part, even if only a minor part, in influencing outcomes. Doyle and McEachern (1998) discuss the concept of power, during which they explain the views of Talcott Parsons (1957)

> For Parsons, power is not just negative or coercive but as he expresses it, a medium like money, one that enhances or increases the capacity to get things done collectively that would be difficult or not possible to achieve individually. (Doyle & McEachern, 1998:26)

The concept of power that will largely be used in this study is that which is used to bring about the agreements necessary to allow this enhanced power of collectivity. Influence perhaps better sums up this notion, rather than power. Influence is also more consistent with the notion of reaching decisions through multi-participatory discourse.

Consensual politics as was mentioned previously is a feature of German domestic politics; it is also necessarily a feature of international regimes. German delegations could be argued to be in a relatively strong position negotiating in such international situations as their political background gives them experience in similar conditions. It will be seen later in this study that during the discussions for the creation of the UNFCCC, representatives of each country tried to negotiate the best possible deal in relation to their own interests. It could be said that this reflects the realist tenets of self-help in an anarchic world order. However, it could also be argued that neo-liberal institutionalism better fits the facts as although each country tried to maximise their own bargaining position, most, if not all States had to compromise, albeit to varying extents. From a Habermasian point of view, discursive processes were engaged in that led to consensual agreement, albeit with a continuing need for discussions aimed at achieving further agreements. Differing viewpoints of States included the following. The USA and other oil producing States were (and in large part still are) primarily concerned with their economic welfare. A large part of the developing world was concerned that the developed world, having itself received the benefit of industrialisation, was trying to prevent developing countries from achieving a better standard of living for their peoples. Small island States are, as has been previously stated, amongst the most vulnerable States (Bangladesh, a low-lying country of continental Asia is an example of an extremely vulnerable non-island country) to the deleterious impacts of climate change and therefore advocated industrial restrictions. Clearly what is considered to be just can be seen differently by different actors. Clayton argues that

> The perceived fairness of an environmental policy will depend on who is defined as a relevant member of the justice community: who is entitled to have his or her interests considered. (Clayton, 2000:472-3)

This statement can be built upon from a Habermasian perspective, by arguing that if all those affected by an environmental policy have their interests considered, i.e. if all are considered to be a part of the justice community, then the resulting policy will be more likely to be perceived as fair.

The signatories of major climate change international agreements are States and the European Community. Realists would argue that the nation-state is the community that is being considered by all States; it would seem that many States, particularly the USA, industrialising States and indeed the Alliance of Small Island States (AOSIS), fight for their own immediate interests. However, other more global notions of citizenship and humanity can be held. Those States that want to industrialise believe that they are just as entitled to, as those States that have already industrialised, i.e. they claim entitlement as part of a global community to be able to fully participate in that community. AOSIS and other environmentally and economically vulnerable States hold that whilst industrially restrictive measures would benefit them in terms of environmental improvements, these policies would also benefit countries worldwide as all will suffer in the long term if no or insufficient mitigation policies are introduced. Germany whilst it may benefit in the long term is looking toward a more global solution. The neo-liberal institutional framework of States working to the benefit of all, fits this more global idea of a community working towards just outcomes for all involved, than does realism. However, I argue that Habermas' discourse ethics is the most appropriate theory to aid evaluation of Germany's climate change international relations. It will be seen that Germany's aims in the international relations of climate change have been influenced by viewpoints from multiple sectors of the German community, as well as by exogenous influences that have been considered and through reflexive thought and discussion been integrated into Germany's policies. It will also be seen that whilst Germany does consider its own strategic interests, its 'actions *cannot be reduced* to this purpose'. (Stokke, 1998:135) Other considerations that Germany takes on board include: the well-being of future generations (as can be seen by its active pursuit for international acceptance of the precautionary principle); and the wider international community, especially those parts of the community that are least able to participate in international negotiations (it actively promotes participation, often funding the attendance at meetings of those unable to fund themselves). It is also the case that Germany conducts its international relations in a manner that can be related to discourse ethics. It will be seen that Germany actively pursues discursive processes in order to try to reach mutual understanding and agreement.

3

EVOLUTION OF CLIMATE CHANGE POLITICS AND POLICIES

Germany's international relations of climate change, the influence that Germany manages to achieve and how it does so are the main purposes of this study. In relation to this statement, scene-setting is the main function of this chapter. To enable assessment of Germany's actions and influence in contemporary international relations, it is necessary to establish the positions of Germany; the European Union, through which Germany often works; and the main international agreements established at UNCED. This chapter will examine the evolution of politics and policies within Germany and of the relevant international agreements emanating from UNCED, thus providing a basis for greater understanding of events analysed in the following chapters. The development of policies within the European Union will be dealt with in chapter five.

Before examining the development of Germany's environmental policies it is pertinent to provide an overview of Germany's political system. Following the Second World War the German political system was completely overhauled. Diffusion of power was a prime objective and this resulted in a system which requires multi-participation and consensual policy-making. Participants in the political system include *Länder;* these are regions that constitute the Federal Republic of Germany. The *Länder* introduce some laws but they implement those made by both themselves and the Federal Government. Implementation is clearly an important factor and can obviously be effected with more or less vigour. Representatives of the *Länder* constitute the *Bundesrat*, which is effectively the upper house of parliament.

Political parties also play a major role in the German political system; Article 21 of the Basic Law states that 'political parties shall participate in the formation of the political will of the people' (Lees, 2005:21). The main political parties, the Christian Democratic Union/Christian Social Union and the Social Democratic Party

> have consciously defined themselves as mass organisations, with relatively large memberships and broad electoral bases, which bridge traditional electoral cleavages, especially class and religion. … Precisely because of their broad appeal, the *Volksparteien* [peoples' parties] must reconcile a wide range of interests within their ranks, (Green & Paterson, 2005:5)

Germany's electoral system is based on proportional representation. A second party list allows for people to express a preference for, in effect, the party that they wish to be in coalition with their main preference party. This system has resulted in coalition governments for all but four years since 1949. These characteristics allow for minority groups to have attention paid to their concerns. However, before gaining representation a political party must receive at least 5 per cent of votes; this it is argued

> limits the number of 'effective' parties within the party system, and isolates extreme Right- and Left-wing parties. (Lees, 2005:21)

The judiciary also plays an important part in German governance. Germany has a written constitution that allows for

> A programmatic statement of general principles …[which are] seen as an essential prologue to legislation and policy development, a tendency that is probably reinforced by the practice of coalition government in which political parties of different ideological persuasions have to come to some agreement on the running of government. Moreover, the emphasis upon constitutionalism in the conduct of government also has the effect of making the policy process more formal. (Weale, 1992:81-82)

A statement of environmental principles (precautionary measures, polluter-pays and industry-government cooperation) was made in 1971. Weale argues that these principles have effectively been institutionalised within the German policy-making system. The institutionalisation of these environmental principles in policy-making is reflective of the linkage between behavioural expectations, communicative action and the law that is identified by Habermas and detailed in chapter two. These environmental principles will be discussed in greater detail in the following section.

Proportional representation and coalition government enables small groups to be heard. It will be seen that when environmental issues including climate change became fairly widespread concerns, the Green Party gained support. This had the effect not only of the Green Party gaining, in the first instance *Länder* government places and later Federal Government positions, but also forcing the main parties to take on board environmental issues. The adoption of people's ideas and concerns by the main political parties was a result of the political system that facilitates such inclusiveness.

In addition to federalism and political parties, Katzenstein (1987) identifies parapublic institutions as being a part of the policy-making process. In the environmental field such institutions include the German Advisory Council on Global Change and the Council for Sustainable Development. Lees (2005) also includes academic research institutes and non-governmental organisations in his discussion on the input of parapublic institutions to the consensual political system. Contributions from such institutions are included in the analysis of the development of the German position for the WSSD in chapter four.

The political features as described above create an environment in which consensual politics can flourish. This propensity for multi-participatory and consensual politics has allowed the ideas and concerns of various sectors of the community to be considered in the development of climate change related environmental politics and policies. As explained previously, multi-participation in the development of policies that are consensually agreed upon are criteria that are consistent with Habermas' theory of discourse ethics. Habermas' work has been conducted in post Second World War Germany and hence may well have been influenced by its political system.

Germany early 1970s to 2002

Environmental politics have become an integral part of both national and global political arenas in recent years. In many instances environmental issues are transnational; hence the national and international are often interlinked in environmental politics. Events at both national and global levels impact upon the other: there is reflexivity. Bearing this in mind, this section will concentrate on the development of environmental awareness and politics in Germany since the 1970s. However, some international links will be included as will a brief background of environmental issues and attitudes prior to the 1970s.

In the post Second World War period, the economic reconstruction of West Germany was a major objective; environmental considerations were not a priority. However, by the 1960s, it was evident that the industrial processes that had enabled economic recovery were having detrimental effects on the environment, this was particularly so in the industrialised region along the river Ruhr. Poor water and air quality in the region were catalysts for action to be taken, whilst improvements in air quality would also reduce the contributions to climate change, this was not a recognised issue at the time.

Although the possibility that human activity could have an effect on the climate was the subject of a paper presented by Hermann Flohn at the University of Würzburg in January 1941 (Cavender & Jäger, 1993:3-18) and of many other papers that Flohn wrote, little attention was paid to them. During the 1950s the possibility of human activity influencing the climate was brought to the fore following the beginning, in 1952, of nuclear weapons testing by the USA in the Marshall Islands. Scientists from around the world were concerned that such activity could affect the world's weather. In Germany, Flohn was called upon to establish a system to measure radioactivity in the atmosphere. After this the possibility of human induced climate change although reported from time to time, did not gain widespread attention until the 1970s.

Forests are important in German mythology and it appears that inherent in German culture is the idea of forests being permanent features of the environment. Beuermann & Burdick (1997) point out that *nachhaltige Forstwirtschaft* (sustainable forestry) was an aim as long ago as the turn of the nineteenth century. It is therefore, not surprising that one issue that engendered popular concern was the widespread discovery in the early 1980s of *Waldsterben* (dying forests), the cause was attributed to air pollution. Von Weizsäcker states

> All of a sudden, the public was up in arms again. The forests, the home of German myths and fairy tales were dying. (Von Weizsäcker, 1994:20)

Beuermann & Burdick argue that due to this entrenched association with forests, during the 1980s and 1990s tropical deforestation and climate change[1] were predominant in German thinking in relation to the concept of sustainability.

Weale states that 'the sudden upsurge in public concern over issues of forest death and damage' (Weale, 1992:1) in 1982, caused the German government to change from a 'cautious stance on the environment' (Ibid) to a more proactive one. It would however, be incorrect to think that environmental issues were not addressed prior to this date. To illustrate this fact there follows a list of some of the environmental laws and acts passed preceding 1982.

Federal Water Resources Act 1957
Federal Emission Control Act 1974
Requirement for Environmental Impact Assessments 1975
Federal Water Law 1976

Effluent Charge Law 1976
Chemicals Law 1980

Weale also points out that environmental functions of government were located in the Interior Ministry in 1969; in 1972 the Council of Environmental Experts was convened; and that in 1974 the Federal Environment Office was created to research requirements for possible regulations.

Prior to the 1970s, environmental policies tended to be reactive and were mainly introduced at the municipal level. During the 1970s there was a shift of responsibility and environmental policy-making became focused at the Federal level and was mainly precautionary.

The UN Conference on the Human Environment was held in Stockholm in 1972. It is certainly possible that at least some of the above activity was in part due to the research, information available and the reflexive thought that would have occurred around that time.

The Laws and Acts detailed above are Federal regulations. Von Weizsäcker notes that

> Modern environmental protection came into being in Northrhine-Westphalia in the 1960s; the first air quality law was adopted in 1963. (Von Weizsäcker, 1994:17)

Von Weizsäcker goes on to say

> But six years later, when Willy Brandt, then leader of the opposition at the Federal level, campaigned for 'blue skies over the Ruhr district', he was ridiculed by his pragmatist political adversaries. Following his election in late 1969, he was the first Federal Chancellor to give high priority to environmental protection in a government programme.
>
> In October 1971 a Federal Environment Programme was adopted, which entailed the passing of eighteen major environmental laws over the subsequent five years, (Ibid).

This was Germany's first Environment Programme. Cavender and Jäger explain that the programme was

> based on three principles – Vorsorge (precautionary measures), Polluter Pays, and Industry-Government Cooperation - … the philosophical underpinnings that have guided all subsequent environmental policies.(Cavender & Jäger, 1993:7)

Each of these principles can be related to principles of justice. Precautionary measures are indicative of considerations of intergenerational justice.

The polluter pays principle demands that the originators of the problem pay to resolve it; this Paterson (2001) argues is a form of retributive justice. Retributive justice can be seen as taking responsibility for distributive injustices and making amends for them. Paterson points out that whilst as an ethical principle retributive justice is largely agreed upon,

> in the climate change context it becomes complicated by the empirical debates concerning responsibility for causing climate change. (Paterson, 2001:121)

This point is valid at both national and global levels, however it is far more complex at a global level as many countries have contributed to greenhouse gas emissions to varying extents, and it is nigh impossible to appoint exact amounts of blame and hence responsibility.

As previously discussed, participatory justice enables ideas to be heard and acted upon, thus enabling policies to be perceived as being just, resulting in acceptance of such policies, which enhances their potential for success. Industry-government cooperation can be seen as a part of this process.

Both the precautionary and polluter-pays principles have been adopted by the international climate change regime (at least they are included in documentation if not properly implemented) since they were introduced in Germany. It can be argued that this transmission of ideas has been achieved through what Habermas terms communicative rationality, in other words the better argument has prevailed, resulting in the above principles entering into international environmental discourse.

As previously stated Willy Brandt placed much importance on environmental protection and it was under his Chancellorship (1969-1974) that the first Environment Programme was introduced. Brandt was leader of the Social Democratic Party (SPD). Helmut Schmidt also of the SPD was Chancellor between 1974 and 1982.

Von Weizsäcker (1994) argues that the policies introduced under Willy Brandt's leadership succeeded in producing cleaner rivers and air in industrial areas such as the Ruhr. However by 1975 the global oil crisis was biting, both economically and in fears of worsening unemployment and continued fuel cost rises. This had the effect of industrial leaders putting pressure on the government not to apportion further costs to industry in the form of environmental regulations. Thus environmental progress slowed in the following few years. Nevertheless, as noted in the preceding list some laws were passed in this time. It is also the case that *Länder* and communities were made to take on environmental administration; emissions standards regarding the protection of water were also introduced. Rowlands (1995) points out that during the oil crisis, although the use of coal increased, energy use in total decreased in West Germany. Economic growth did not suffer as perhaps would be thought, instead energy use became more efficient. Rowlands attributes this fact with the reason why German businesses were not violently opposed to proposed policies to reduce CO_2 emissions in the late 1980s; he claims that instead these policies were seen as being potentially beneficial in terms of efficiency and productivity.

The need for emissions standards to reduce and control air pollution came to the fore when the issue of *Waldsterben* was discovered and made public. This realisation was the catalyst that reinvigorated progress in German environmental policies. An example of this reactivation is given by Weale (1992), who comments that following the German government's change in 1982 to a proactive stance in its environmental outlook, in 1983 strict emissions limits with regard to sulphur dioxide and large furnaces were implemented under the Federal Emission Act. Weale goes on to say that Germany has pursued environmental policies such as the restriction of vehicle emissions; waste recycling programmes; and waste disposal regulations.

In 1982 the government changed, a coalition was formed of the Christian Democrats (CDU), the Christian-Social Union (CSU), and the Free Democratic Party (FDP). The new Chancellor was Helmut Kohl of the CDU; he remained Chancellor until 1998. As a result of public concern over *Waldsterben* the new government were vociferously called upon to take an active stance regarding environmental policies. It is worth noting that members of the Green Party (which had formed from a social movement that had arisen in the 1970s due to popular concern over the environment and particularly over concern about nuclear safety[2]) were voted in to the *Bundestag* (lower house of parliament) in 1983. At the level of the *Länder,* the Green Party was successful in gaining representation in Baden-Württemberg in 1980; in Hamburg, Lower Saxony, and Hesse in 1982; and in Bremen in 1983. In Hesse following the 1985 state elections the Green Party and the SPD entered into a coalition and Joschka Fischer

> became the first Green state minister for environmental and energy affairs. (Mewes, 1998:41)

These events are significant as they demonstrate that 'green' issues were a matter of popular concern; this in turn put pressure on other political parties to pay attention to environmental matters. An example of the results of this pressure is the adoption by the SPD during the 1980s of environmental policies that were in large part those already espoused by the Green Party. Lees argues that this was done 'to counter the electoral challenge from the Greens' (Lees, 2002:10). The electoral success of the Green Party and the adoption of environmental polices by mainstream political parties is an example of multi-participatory and inclusive processes that can be related to Habermasian decision-making procedures.

Cavender and Jäger (1993) explain that environmental non-governmental organisations were relatively (to those in other countries) late in campaigning at a national level and with regard to climate change issues. They attribute this mainly to the opposition of these organisations to nuclear energy policies within Germany. The nuclear energy industry had played on the climate change issue to promote their industry and the CDU used this argument in the 1980s to justify the planned expansion of nuclear energy. Environmental non-governmental organisations did not want to appear to be vicariously supporting this argument by campaigning on climate change issues. Following the Chernobyl nuclear accident on 26 April 1986 much attention was given to the questionable safety of nuclear energy. Hatch (1995) reports that along with growing public concern, there were party political debates regarding the possibility of ceasing nuclear power production. The Green Party advocated immediate closure of all nuclear power plants; the January 1987 election saw a substantial swing (from 5.6 per cent in 1983 to 8.3 per cent) toward the Green party (Hatch, 1985:421). Clearly, there was public support for environmental consideration, including the proposed withdrawal from reliance on nuclear energy. Post Chernobyl the SPD set up a commission to investigate the viability of a non nuclear energy powered Germany; this found that it would be possible if nuclear sourced energy were phased out over a period of ten years by using coal instead. In the third quarter of 1986 the SPD adopted this stance on energy. Burning coal to produce energy contributes to global warming and as mentioned above the CDU used such facts to promote nuclear expansion.

Hatch (1995) points out that the Interior Ministry, which was responsible for the environment mishandled events post Chernobyl and that this resulted in the creation in 1986 of the Federal Ministry for the Environment, Nature Conservation and Nuclear Safety (BMU). He also points out that the Ministry of Transport handled meteorological issues and hence controlled climate change issues until late 1988. The Ministry of Transport failed to provide for adequate participation in the initial negotiations of the Intergovernmental Panel on Climate Change (IPCC), this led to responsibility for climate change being transferred to the BMU.

Weale (1992) argues that during the 1980s Germany became the environmental leader within Europe and that Germany's reunification in

1990 reinforced the idea of the need for environmentally protective action, as East Germany's dire environmental situation was realised. Weale points out that

> In regard to problems of global climate change and the control of greenhouse gases, Germany had by the beginning of the 1990s committed itself to the ambitious target of a unilateral reduction in carbon dioxide emissions of between twenty five and thirty per cent by the year 2005. (Weale, 1992:69)

It could be argued that once the German government became aware of the extreme effects of environmental degradation in East Germany, they were motivated to act in a just and fair way, i.e. to try to ensure that all people, both present and future are not forced to suffer the grim consequences of unsustainable use of the environment.

Beuermann and Jäger (1996) point out that climate change politics and issues such as recession are interlinked. The former East Germany can be used to illustrate this point as following reunification their economy suffered greatly, the massive reduction in industrial outputs clearly contributed substantially towards the decrease in their CO_2 emissions by 14 per cent between 1987 and 1992 (Beuermann & Jäger, 1996:206). The slowing of environmental policy-making during the oil crisis also demonstrates such linkages.

Beuermann and Jäger generally seem to be far more critical of Germany's contribution to climate change politics and indeed to its contribution to climate change itself. They state that 'The per capita CO_2 emissions in Germany are about three times the global average and about twice the average for the EU countries as a whole' (Beuermann & Jäger, 1996:191). It is not stated when this comparison was made, however most of the other statistics given on that page are for between 1987 and 1993. The book was published in 1996 so it is likely that the comparison was made in the early 1990s. It is also worth pointing out that especially with regards to the global comparison, like is not being compared with like. It would perhaps be more informative to compare Germany's emissions with other countries within the Organisation for Economic Cooperation and Development (OECD). It is also a static comparison; it would perhaps be more informative to give a number of comparisons over time so that Germany's efforts towards improvement could be assessed vis-à-vis the progress of other similar countries. The OECD Environmental Data Compendium 2002 details per capita CO_2 emissions in the late 1990s,[3] a selection of which are: Germany 60.3 kg, EU15 93.4, OECD Europe 87.1, OECD 152.8, France 136.2, UK 80.0. Beuermann and Jäger (1996) stated that per capita CO_2 emissions in Germany were about twice the average for EU countries. As the more recent data from the OECD shows, by the end of the 1990s Germany's per capita CO_2 emissions were approximately two thirds of the EU average. This is a positive indication of the success of Germany's efforts to reduce emissions.

Somewhat contrarily Weale, and Beuermann and Jäger seem to differ in their assessment of the German political system in an opposing way to the way one would think. Weale states that 'With a federal system of government, players in the policy game usually have less room for manoeuvre than actors within a unitary system' (Weale, 1992:55). Beuermann and Jäger acknowledge the complexity of the bargaining system within the federal system; but they also make the point that federalism combined with the political party system (presumably here they mean proportional representation) and consensual politics allows for public opinion to be quickly taken on board and hence acted upon. Thus media coverage of such eventualities as *Waldsterben,* depletion of the ozone layer and global warming are important as they raise public awareness of such issues. If the results of scientific research into environmental issues can achieve media attention, this too, clearly feeds into the political system. Scientific research is not however, reliant on media attention as Germany has a clear structure for scientific input into the political system. Mentioned above was the formation of the Council of Environmental Experts in 1972. Other examples include: the Federal Environment Agency *(Umweltbundesamt - UBA)*

> which was established on 22 July 1974, … to provide scientific and technical support for the Federal Environment Ministry (www.bmu.de/english/tasks/uba.htm October 2001)

and the German Advisory Council on Global Change (WBGU) which was established by the Federal Government in 1992. The institutionalised inclusion of scientific experts in the policy-making debate is another indication that multi-participative Habermasian style decision-making procedures are in place.

Beuermann and Jäger, and Beuermann and Burdick also mention the input of Enquette Commissions. These are explained as

> an advisory body to the federal government. One half of the members are politicians, the other are scientists appointed by parties represented in parliament. Enquette Commissions are established to give policy advice on complex political issues. (Beuermann & Burdick, 1997:102)

Beuermann and Jäger hold that the work of the Enquette Commission *Vorsorge zum Schutz der Erdatmosphäre* (Preventive measures to protect the Earth's atmosphere), which was set up in October 1987 was successful and shortened the length of time the political process would have taken without such input. Following the work of the Enquette Commission and the Federal Environment Agency

> On 11 December 1991 the Cabinet … stated that Germany would aim to reduce CO_2 emissions by 25-30 per cent by the year 2005 based on 1987 values. The Enquette Commission in its 1990 report had a goal of 30 per cent CO_2 emissions reduction. …
>
> As a further reaction to the final report of the first Enquette Commission in September 1991, the German Bundestag agreed strictly to apply the precautionary and polluter pay principles and to integrate environmental protection in all political areas (BT.-Drs. 12/1136). This application is stressed in every environmental resolution because these principles are the basis of German Environmental Policy since 1971. They are used to explain why Germany puts emphasis on the limitation of greenhouse gases and not on adaptation research and measures. (Beuermann & Jäger, 1996:195)

As previously noted one of the principles enshrined in the 1971 Environment Programme is that of industry-government cooperation. The consequence of such cooperation is, unsurprisingly, contested. The Directorate General XI, Environment, Nuclear Safety and Civil Protection of the Commission of the European Communities (1993) hold that close cooperation between the various levels of the German government and industrial associations is a positive factor in environmental policy-making, as is the involvement of environmental non-governmental organisations. Examples of such are *Bundesverband der deutschen Industrie* (BDI – Federal Association for German Industry) and *Bund fur Umwelt und Naturschutz Deutschland* (BUND - German Federation for Environment and Nature Protection). The inclusion of industrial associations and environmental non-governmental organisations in discussions with the German government are further indicators of the existence of Habermasian style discursive decision-making processes. However, Beuermann and Jäger point out that the BMU is one of the smallest and resource poor ministries. The Federal Ministry for Economics and Labour (BMWA), the Federal Ministry for Agriculture, Nutrition and Forestry and the Ministry of Transport all make decisions that affect climate change, however they not only represent

> the German government, [they] also take into account the interests of industry and others. BMU does not have such connections but stands between all lines: other ministries, industry and environmental groups. (Beuermann & Jäger, 1996:195)

They point out that industrial influence with various ministries and inter-ministerial conflict has the effect of weakening the environmental policies of the BMU. Alternatively, it could be argued that such governmental workings are demonstrative of the consensus oriented nature of German politics and comply with Habermas' contention that all parties should be able to partake in discourse.

Events and ideas have been highlighted that have put pressure on the German government to follow certain domestic and foreign policies. Values and ideas that were evidently held by a large proportion of the German populace were also those held by Klaus Töpfer a member of the CDU and Environment Minister between May 1987 and November 1994.[4] According to Von Weizsäcker he

> soon brought both national and international recognition to the ministry. Helped by the Commission [European] and by France, Denmark and the Netherlands, he pushed the Large Combustion Plants Directive through the EC Council... He also took a leading role internationally on such topics as protection of the North Sea, the rain forests, global warming and the ecological reconstruction of Eastern Europe. At the Earth Summit in Rio de Janeiro he was certainly one of the leading figures representing the North and contributed much to the ultimate success of UNCED. (Von Weizsäcker, 1994:20)

Beuermann and Burdick agree with this positive opinion of Töpfer they state that

> Minister Töpfer was responsible for the strong positions and political leadership of Germany in climate change politics in the end of the 1980s and the early 1990s. (Beuermann & Burdick, 1997:105)

They also argue that a progressive environmental policy was developed at the Federal Ministry for Regional Planning, Building & Urban Development after Töpfer moved there in November 1994.

Beuermann and Jäger point out that whether or not the detrimental affects on the environment are actively considered and acted upon also depends upon the importance placed on other issues at varying times.

They use as an example the reunification of East and West Germany when many politicians and the Ministry of Transport thought that although

> environmental considerations which by law are part of the planning procedures, [they] should be given lower priority until the most urgent infrastructure projects have been implemented. (Beuermann & Jäger, 1996:210)

In effect Beuermann and Jäger argue that whilst Germany does pay some attention to climate change issues, they are given less priority than more entrenched industrial visions of development. Sturm (1996) expresses this dichotomy in more neutral terms, although in essence he points to the same results. He cites the decision of Federal Government in 1995 to plan to introduce car taxes based on emission levels; this apparently coincided with the car industry's wishes. The Environment Minister, Angela Merkel wanted engine capacity to be the taxation criteria. The environment is effected by industrial processes and conversely measures to protect the environment impact on industry. These often competing interests are evident in policy-making discussions. The Habermasian position is that when true consensus is not achievable, compromise is needed. It can be argued that the situations detailed above are examples of issues where compromise has been necessary.

Following the election on 27 September 1998 a Red-Green coalition formed the government led by Chancellor Gerhard Schröder. Jürgen Trittin of the Green Party was appointed as the Environment Minister; Joschka Fischer also of the Green Party was made Foreign Minister. Another ministry that is pertinent to this thesis is that of Economic Cooperation and Development, the Minister since 1998 has been Heidemarie Wieczorek-Zeul who is a representative of the SPD (the 'red' part of the coalition).[5] Both the SPD and the Green Party are in favour of phasing out nuclear power and this was one of their first environmental goals once they had formed a government. Lees (2005) explains that following lengthy negotiations with a variety of interested parties, the phasing out of nuclear power in Germany was finally agreed in June 2000.

Clearly it was a major intention of, at least the Green Party if not the whole government to make their mark on Germany's environmental politics. In their first two years of government a number of measures were put in place to help reduce CO_2 emissions. These were:

- The ecological tax reform, which envisages a gradual increase in energy prices in all segments in order to create incentives for the development and market launch of new technologies, coupled with the rational and economical use of energy.

> • The Renewable Energy Act, which promotes the conversion of renewable energy sources into electricity.
>
> • The market launch programme for renewable energy sources, which particularly benefits the use of solar panels, as well as rational energy use.
>
> • The "100,000" roofs programme, which supports investments in photovoltaic systems.
>
> • The promotion of low-sulphur and non-sulphur fuels also helps to achieve a breakthrough in fuel-efficient, low-emission engine technology. (BMU, 2000:7)

The BMU proposed a new climate change protection programme; this was adopted on 18 October 2000. The summary of the programme published by the BMU states that

> The Federal Government's targets are extremely ambitious:
>
> • To reduce emissions of carbon dioxide by 25% by 2005 compared with 1990 levels.
>
> • To reduce emissions of the six greenhouse gases cited in the Kyoto Protocol by 21% between 2008 and 2012, within the context of EU burden-sharing. 1990 is the base year for CO_2, CH_4, and N_2O, and 1995 for H-CFC, CFC and SF_6.[6]
>
> Furthermore, for the first time the Federal Government has also set up specific targets relating to technology and energy resources:
>
> • A doubling in the proportion of renewable energy sources by 2010 compared with current levels, and a further substantial increase in the proportion of renewable energy sources after 2010.
>
> • The expansion of combined heat and power generation by means of set quotas, aimed at cutting CO_2 emissions by an additional 10 million tones by 2005, and by 23 million tones by 2010.
>
> • A significant increase in energy productivity over the next few years.
>
> Finally, for the first time, the Federal Government has set reduction targets for individual sectors. (BMU, 2000: 4)

Before going on to detail the policies that will be introduced to enable the above targets to be met the summary makes the following point

> Although the previous Federal Government set ambitious climate protection targets, it failed to take effective action which would have enabled it to meet these targets. (BMU, 2000:5)

It is stated that 'Present shortfalls create a difficult starting position' (Ibid). The results of actions taken by the previous government are then detailed. The above quote reflects the fact that there was no major change in objectives in climate change politics when the government changed. It can be argued that such continuity of objectives may well be due to the multi-participative and consensual nature of politics in Germany. New policies that the Red-Green government introduced are new instruments rather than new objectives. The summary goes on to describe the policies being introduced. It is not within the remit of this study to reproduce verbatim the detail given, it will suffice to list the sub-headings: Expansion of combined heat and power generation; Adoption of the energy saving ordinance; Subsidy programme to reduce CO_2 in existing buildings; Declaration by German industry on climate protection; Package of measures for the transport sector; Creation of an additional working party to the interministerial working group on CO_2 reduction; Voluntary commitment by the Federal Government; Other greenhouse gases.

More overarching intentions and beliefs are then explained, such as the importance of EU and international negotiations and policies. Dedication to the continuance of climate change protection policy at both national and international levels is identified, as is the belief that the commitments of industrialised countries under the Kyoto Protocol's initial commitment period should be increased in future phases. It is believed that developing countries, especially those with increasing emissions should be included in future commitment periods. However, a commitment to the continuance of support for climate change protection polices in developing countries is made, commitments will be reviewed and updated. It will be seen later in this study that the proposal to include developing countries in future commitments periods proved to be controversial.

The climate protection programme summary also states that

> The Federal Government also welcomes the achievements of the Länder, cities and communities, environmental and consumer organizations, trade unions and other social organizations, and calls on these players to remain vigilant in their endeavours.
>
> ...
>
> The Federal Government sees the adoption of a national climate protection programme as a vital pre-requisite of sustainable development and the national sustainability strategy. (BMU, 2000:11)

This statement acknowledges the importance of multi-stakeholder participation in order for climate change policies to be effective. The national sustainability strategy and its formation will be examined in the following chapter.

Mentioned above is the German Industry declaration on climate change; following on from this, an 'Agreement on Climate Protection between the Government of the Federal Republic of Germany and German Business' was reached on 9 November 2000. In effect it was agreed that business and government would work in cooperation toward meeting the goals of the Kyoto Protocol. Business agreed to increase the emissions reductions it had previously agreed to introduce; all six greenhouse gases would be targeted with the aim of total reductions of 35 per cent by 2012 on 1990 levels, with the intention that CO_2 emissions be reduced by 28 per cent. Calculations used to arrive at these figures were those used by a jointly authorised neutral monitor. The voluntary commitment by German industry relates to energy per unit of GDP and not to absolute reductions.

For its part, the German government agreed to take into consideration German business when developing further ecological tax reforms and to try to ensure that business's international competitiveness would not suffer as a result of climate change related policies.

The agreement between government and business is to be monitored by an independent scientific institute as has previous agreements; the costs of which will be born jointly and equally by the government in the form of the BMU and German business. The agreement was signed by: The Federal Chancellor, Gerhard Schröder; Federal Minister of Environment, Nature Conservation and Nuclear Safety, Jürgen Trittin; Federal Minister of Economics and Technology, Werner Müller; President of Federation of German Industries; Dr. Hans-Olaf Henkel; Vice President of Federal Association of the German Gas and Water Industries, Dr. Erich Deppe; President of Association of German Electricity Supply Companies, Günter Marquis; and Chairman of Federation of Industrial Energy Consumers and Self-Producers. Accountability and transparency appear to be in evidence; this could be explained as endeavouring to show that agreements are carried out justly.

Previously in this section it has been pointed out that Beuermann and Jäger are amongst those people that argue that economic and industrial interests are entrenched within German politics and hence tend to outweigh environmental concerns. However in November 2001 the German government agreed to increase renewable energy subsidies for the following four years; this was despite the fact that Werner Müller, the Minister for Economics wanted to cut financial support. According to a Planet Ark report dated 19 November 2001, Müller wanted to reduce support by DM 100 million (from 300 million) and in fact it was raised by DM 100 million. The government's renewable energy research budget was cut from DM 300 million to DM 274 million; however Müller had wanted this cut to DM 235 million. The report also notes that the Science Ministry may provide another DM 6 million (Planet Ark, 19-11-01). It is evident that whilst economic and industrial interests are necessarily considered, these interests are tempered by environmental concerns. Dialogue between the various parties allows for consensual agreement to be reached. Multiple viewpoints are put forward and considered; this could be seen as at least part way towards Habermas' ideal of arriving at consensus through discursive processes the result of which is determined by the force of the better argument. Whilst this ideal was not fulfilled, a consensually agreed upon compromise was achieved.

In January 2002 the BBC reported that Germany had over 11,000 wind turbines and is the world's leader in wind power production, in 2001 production capacity increased by 44 per cent. The German government has agreed to phase out nuclear energy and is investing heavily in wind power. It is also stated that considerations are under way

> to build up to 5,000 wind turbines off Germany's north coast. Some would be located in open sea up to 45 kilometres (27 miles) offshore, a feat never before attempted. Since the wind is stronger at sea, the energy potential is highly attractive. Giant wind turbines, double the sizes of conventional ones, are being developed for offshore use. A pilot project has already been authorised and is expected to be operational next year, (BBC, 16-01-02)

The expansion of solar power was also an aim of the German government. In a report on a speech given by Jürgen Trittin in March 2002 at a conference on solar thermic electricity generation, it is asserted that

> Germany's research institutes and companies are world leaders in the development of solar power stations …[and that] "With the Government's help, solar power stations will become Germany's export hit of the future. This will benefit both climate protection and the economy."
>
> Under its investment in the future programme, the German Government will provide a total of 10 million Euro up to 2003 for research and development projects on solar thermal power stations. … The Federal Environment Ministry has invested a total of 30 million Euro in renewable energies research. Jürgen Trittin believes that the investment in the future programme should be continued until at least 2006. (BMU, 14-03-02)

CO_2 emissions from vehicle emissions are a major contribution to GHGs. The German Environmental Report 2002 states that road traffic accounts for 52 per cent of the country's CO_2 emissions and that these emissions rose during the 1990s, but that since 1999 they have begun to decline. This trend is due to the reduction of fuel consumption; since 1990 the average car's fuel consumption has reduced by 17.8 per cent. The same report states that the government aims to reduce freight traffic by 5 per cent by 2020 on 1999 levels and by 20 per cent for passenger traffic. In 2001 approximately the same amount of funding was given to the railways as was given for road construction; other policies that have been introduced with the aim of reaching these targets are pilot projects involving environmentally friendly buses in Berlin and Frankfurt/Oder and the introduction in 2002 of a national cycling plan. Since 1999 the government has introduced ecological tax reforms. Elements of this follow:

> The mineral oil tax on fuel has risen by 3.07 cents per litre annually. The ecotax provides an incentive to purchase fuel-efficient motor vehicles and to drive economically… owners of natural gas-powered vehicles will only pay a significantly reduced mineral tax until 2009…
>
> In 2001, the "flat mileage rate" in Germany was abolished and replaced by a "flat distance rate" which is the same for all modes of transport. The flat-rate will encourage employees to switch to other modes of transport such as the railways or cycles.
>
> From 2003 onwards the Federal Government will be introducing a distance and emissions-based motorway toll for heavy goods vehicles… (BMU, 2002)

The aim is to encourage people to use public transport and freight to use trains and waterways.

There were disagreements in the German polity about the introduction of heavy goods vehicle tolls. The law allowing for tolls was passed by the government in August 2001; however the German parliament did not approve the law until March 2002. *Länder* representatives had stalled the passing of the law through the *Bundesrat* as they wanted variable tolls depending on time and place and for municipal vehicles to be exempted. Agreement allows

> for charges averaging euros 0.15 per kilometre for domestic and foreign lorries over 12 tonnes.
>
> …

> A transport ministry spokesperson told Environment Daily that there were no more political hurdles to entry into force on 1 January 2003 of the law… (Environment Daily, 26/03/02)

It is clear that Germany is taking environmental issues seriously, is acting on them and is also keen to promote energy saving technologies. As Germany is taking an active lead in renewable energy production it is likely to be able to benefit economically from technology transfers, whilst simultaneously enabling the world's environment to benefit.

Although Germany and the EU are often seen as leaders in the battle against climate change, Germany's policies and attitudes do not always coincide with those of the EU. An example of this is the fact that the European Commission investigated the legality of Germany's green energy support laws, on 22 May 2002 they ruled

> that they do not constitute state aid…
>
> Germany's current green energy "feed-in" rules took effect for renewables in April 2000, and for combined heat and power (CHP) in May 2000. Each obliges utilities to connect such installations to the grid and to purchase their power at a guaranteed price above market levels. (Environment Daily, 22/05/02)

It is worth ending this section by noting that the Worldwatch Institute in its 'State of the World 2002'

> confirms that the German government is setting an international example by moving forward with its policy for meeting the Kyoto Protocol targets. (BMU, 15-01-02)

The BMU reports that

> In the renewable energies sector alone, over 50,000 jobs have been created in Germany – a trend that will continue in the future: estimates by notable institutes show that reducing CO_2 emissions by 40% up to the year 2020 is not only feasible but would also create 200,000 jobs. (Ibid)

Summary

This section has examined the evolution of climate change related policies within Germany. It has been seen that there have been a multiplicity of factors that have influenced the evolution of climate change politics and policies in Germany and that discursive inclusiveness has allowed ideas and values to be taken on board. It can be argued that Germany's federal system allows for multiple viewpoints to be considered and that this has resulted in 'green' issues becoming mainstream concerns. This can be related to Habermas' ideal of multi-participative discourse being engaged in that results in consensus being reached through the force of the better argument. It has also been argued that the enshrining in law of the environmental principles contained in the 1971 Environment Programme, i.e. precautionary, polluter-pays and industry-government cooperation, is akin to the process identified by Habermas in which generalisable interests become universal norms. The adoption of environmental concerns by mainstream political parties has led to effective domestic environmental policy-making, which has subsequently allowed Germany to become a leader in climate change politics. The following chapters will support this leadership claim. It is a distinct possibility that Germany has become a leader in climate change politics, in part, because it has been allowed to. The impression gleaned from numerous interviews is that a strong foreign policy in 'soft' issues was acceptable, both within Germany and in the international arena; this may not have been the case with traditional 'hard' issues.

United Nations Conference on Environment and Development

As previously explained the main function of this chapter is one of scene-setting. The purpose of this section is to provide background information about international agreements that will aid understanding of, and be built upon during, analysis of the WSSD and COP8 later in this study. Whilst Germany's contributions to UNCED will be examined, Germany is not the main focus of this section.

UNCED was held on 3–14 June 1992 in Rio de Janeiro. The ministries that were responsible for Germany's contribution to the conference were the Ministry for the Environment and the Ministry for Economic Cooperation and Development. UNCED was arguably the first truly global conference as it was the subject of a wide variety of global inputs and it attracted a great deal of interest worldwide.[7] One hundred and seventy eight countries sent delegations to UNCED, more than one hundred heads of State attended meetings there, along with representatives of more than one thousand non-governmental organisations (NGOs).[8]

The Rio Declaration; The Biodiversity Treaty; The Statement of Forest Principles; *Agenda 21*; and *The Framework Convention on Climate Change* are the five major agreements that are associated with UNCED. The italicised agreements are those that are relevant to this study. During the two years prior to UNCED, four preparatory committees met to draw up the basis for the texts of the first four of the above agreements. The Framework Convention on Climate Change was negotiated separately.

United Nations Framework Convention on Climate Change

The United Nations Framework Convention on Climate Change (UNFCCC) was adopted on 9 May 1992 in New York and was opened for signature at UNCED. As the name suggests it is a framework within which continuing negotiations take place. The essence of the framework is to work toward the

> stabilization of greenhouse gas concentrations in the atmosphere at a level that would prevent dangerous anthropogenic (man-made) interference with the climate system. Such a level should be achieved within a time-frame sufficient to allow ecosystems to adapt naturally to climate change, to ensure that food production is not threatened and to enable economic development to proceed in a sustainable manner. (UNFCCC, climate information sheet 18)

The above statement shows that consensus on the objective of the UNFCCC has been achieved. However, this does not mean that consensus has been found on the criteria for these objectives. For example, agreement has not been reached on what a dangerous level of anthropogenic interference is, or on a sufficient time-frame for eco-systems to be able to adapt. There is however, continuing discourse to try to reach a consensus on these issues.

As was noted with the issue of *Waldsterben,* mass public concern forces issues on to the political agenda; to quote Maddock 'Rosenau argues that popular access to information is one of the most significant modern trends (Rosenau, 1995) and this is as true for environmental governance as for any other aspect of civil society' (Maddock, 1998:245). Climate change negotiations were on the political agenda due to a rise in the mid 1980s of scientific evidence of global warming and the subsequent rise in public concern once this information had been disseminated. In 1988 the World Meteorological Organisation (WMO) and the United Nations Environment Programme (UNEP) established the IPCC. Also in 1988 following a proposal by the Maltese government, the UN General Assembly (UNGA) passed resolution 43/53 on the 'Protection of global climate for present and future generations of mankind'. In 1990 the IPCC published its First Assessment Report in which it concluded that climate change was indeed a threat and recommended that a global treaty be sought on the matter. Later in 1990 ministers at the Second World Climate Conference[9] also called for negotiations on such a treaty to commence. In December 1990 the UNGA adopted resolution 45/212, this initiated negotiations by establishing the Intergovernmental Negotiating Committee (INC) for the Framework Convention on Climate Change (FCCC). It is evident that as a result of the dissemination of knowledge (discourse), it was determined that various international bodies needed to be established in order to work toward a global agreement regarding climate change management.

The INC/FCCC first met in June 1991 at which time it set up two working groups:

> Working Group I – Commitments (to limit and reduce the emissions of CO2 and other greenhouse gases (GHGs); protect and enhance sinks and reservoirs; provide adequate financial resources to enable developing countries to meet incremental costs; facilitate the transfer of technology on a fair and equitable basis; and address the special needs of developing countries); and Working Group II – Institutional and Legal Mechanisms. (Halpern, 1992:7)

The INC met every few months thereafter, however, agreement was hard to achieve as many different viewpoints were held. Broadly speaking the USA differed in their stance from the rest of what is commonly known as the North, the USA were and still are, far more hard-line against industrially restrictive agreements. Germany was a lot more progressive in its views. Views form the South ranged from the hard stance of the oil producing States, to a great deal of the developing world wanting to protect their rights to development by industrialisation, through to the small island States who wanted far reaching curbs on GHGs and a generally strong line taken against the anthropogenic causes of global warming. Thirty five of the worlds small island States had, during the Second World Climate Conference decided to join together in their quest for their concerns, primarily about climate change and sea level rise, to be heard, thus the AOSIS was formed.[10] AOSIS was active in INC/FCCC negotiations and pushed for the following principles that are included in the UNFCCC:

- the principle of preventive action;
- the precautionary principle;
- the polluter pays principle and State responsibility;
- duty to cooperate;
- equity;
- the principle of common but differentiated responsibility, and

- commitment to binding energy conservation and the development of renewable energy sources.
(www.sidsnet.org/aosis/background.htm November 2001)

These principles largely coincide with those supported by Germany and the inclusion of them in the agreement would have been due to consensus being reached by all involved in the negotiations and due to pressure from the delegations of many countries. It is worth noting that the precautionary principle which is seemingly ubiquitous in environmental discourse and modern environmental treaties was first introduced in Germany (Von Seht & Ott, 2000: 7; www.mep-online.com/chapter2/section_2_9_ 10.html September 2004). During the INCs second session a number of States submitted draft conventions (Ulbert, 2003:19-21). Germany's draft included the precautionary principle, binding reduction targets, and the call for immediate action. The USA were against targets and 'doubted the usefulness of the precautionary principle on the basis of the lack of scientific certainty' (Ulbert, 2003:21). The precautionary principle was included in the UNFCCC; this can be seen as the result of discourse in which Habermasian communicative rationality prevailed.

In general, differences of opinion were great and so to a large extent the lowest common denominator appears to have been sought. At the December 1991 meeting it was agreed that despite previous calls (for example in the first IPCC report) for a reduction in CO_2 emissions, that stabilisation, not reduction, targets be sought. The European Community called for the stabilisation of CO_2 at 1990 levels by 2000 (this goal had been included in Germany's draft convention), most delegations agreed with this proposal with the notable exception of the USA. A matter for debate was also whether CO_2 emissions alone should be included in the convention or whether all GHGs should be included.

Another area of dispute related to the call for financial resources and the transfer of technology. The developing world agreed that provision should be made for financial assistance and technology transfers. The European Community recognised a need to provide additional resources but did not make any specific commitments. It should be noted that when the European Community's position is commented upon it is a common position of Member States, which includes Germany. The USA wanted the international community to look into the prospects of multilateral and bilateral funding. Brazil, China and India were interested in the idea that commitments should be made to 'pay for incremental costs of measures to prevent climate change, i.e., to limit emissions of GHGs or to maintain and enhance sinks' (Halpern, 1995:7). AOSIS favoured the idea that the financial resources 'pay for adverse costs of climate change should it occur' (Ibid). Again the economic and environmental welfare/vulnerabilities of various States patently underscored the stance their representatives took. Countries that constitute AOSIS are environmentally vulnerable, they tend to be economically vulnerable too and they do not emit vast amounts of GHGs. AOSIS suggested that an insurance fund be set up so that should global warming prove to be detrimental, recompense could be made.

Cooperation regarding the transfer of technology was seen as being possible along the traditional channels of bilateral or multilateral assistance. The convention also established the Global Environment Facility (GEF) and Activities Implemented Jointly (AIJ). The GEF is government funded and aims to support development of, and demonstration of economically efficient technology that will reduce GHG emissions. The GEF should also promote sustainable development in developing countries, if a technology is seen to be cost and energy efficient, the GEF should pay the difference between the costs of introducing climate friendly technology over polluting technology.

> The GEF was first officially proposed by French and German representatives during the September 1989 meeting of the Worlds Bank's Development Committee. (Rowlands, 1995:196)

AIJ aims to attract the private sector to invest in the transfer of technology and technological knowledge to developing countries.

At the INCs fifth meeting in May 1992 it was agreed that the convention cover all GHGs and that the timetable for stabilisation be 'sufficient to allow ecosystems to adapt naturally' (Ibid). It was also agreed that as developed countries emit most GHGs that they should lead in the effort to combat climate change. Countries would need to submit inventories of sources and sinks of GHGs, and reports on policies to limit sources and improve sinks.

As has already been mentioned it is apparent that each country's delegation fought for what they perceived to be in their own country's best interests, be that economic or environmental, or taking economics and environment into account but with one element taking precedence. The more economically powerful countries certainly managed to limit drastic enforcement of any far reaching measures to combat climate change, however, they were not able to have the convention written purely on their terms. However watered down the restrictions are from a small island State point of view, the aim toward restrictions on GHG emissions is written into the convention. Simplistically it would appear that the fact that each State was working toward its own best benefit fits into the realist framework of an anarchical world order with every State fighting its own corner. The fact that the more powerful States and predominantly the hegemonic USA prevented items being included in the convention that would be too damaging to their economies also fits with realist theory. However negotiations between a variety of actors with plural interests did occur and continue to do so and hence it could be argued that neo-liberal institutionalism can be understood to best describe the situation in which negotiations occurred and continue to occur. However, from a Habermasian point of view both realist and neo liberal institutionalist theories are part of a construction that needs to be changed through discursive processes. It can also be argued that through discursive processes plural interests were considered and to some extent, were taken on board. However, these varying opinions appear to have been heeded to varying degrees that appear largely to correlate with the economic world order. A quote from Clark seems to describe this point aptly

> Diplomacy and bargaining are to international relations what the free market is to economics. Every State sets out to obtain the best deal for itself and its success or lack of it in this effort is largely determined by the resources that it commands. (Clark, 1989:15)

Clark writes from a realist stance in which in a world of sovereign States the order is on the face of it anarchic, though actually States form a hierarchical structure determined by the power that each can draw upon. According to Clark 'every state sets out to obtain the best deal for itself'; this may be the case but it is not necessarily the case that all States or all people think purely in economic terms. Consideration of economic factors may be a necessity but other factors such as the environment, justice, human rights etc. may also be a part of the equation. As O'Hara argues

> Humans are more than rational choice decision-makers. They are moral agents embedded in social and ecological contexts. (O'Hara, 1996:100)

This statement fits with the thoughts of Habermas as explained in the quote by Stokke (1998) that is reproduced in the introduction to this study.

Returning to the specifics of the IPCC negotiations, Paterson's interpretation of the process is that the agenda on global warming was primarily set by the epistemic community 'without any significant entrenchment in national bureaucracies' (Paterson, 1996:147). He explains that although the IPCC delegates were mainly nominated by governments, many were not government members, many were academics. Paterson goes on to point out though that whilst the IPCC mainly consisted of scientists, the INC delegates from developed countries were 'dominated by foreign ministries' (Ibid). Thus initially scientists were influential in putting forward a largely, country neutral debate regarding climate change, however, once it got to the stage that States would have to commit funds or to actions, State politics came much more into play. The fact that scientists looking at climate change from a global viewpoint were able to get their findings on to the global agenda, could be argued to be the reason that those States lower down in the realist political global hierarchy, for example small island developing States (SIDS), were able to have their say in global negotiations. Once climate change was on the global agenda it was up to SIDS, as it was to all countries, to find ways in which they could most effectively negotiate. However, the process just described can be related to Habermas' discourse ethics; all interested parties were able to contribute their opinion in a rational discussion that led to agreement over the form the UNFCCC would take. Participatory justice led to an agreement that was largely perceived as just.

As previously stated the UNFCCC was adopted on 9 May 1992, it opened for signature during UNCED and entered into force on 21 March 1994 having been ratified by fifty States. The INC continued to meet until the First Conference of the Parties (parties to the convention), which was held in April 1995.

At each Conference of the Parties (COP) various interest groups tend to present a largely unified viewpoint. The European Union is one such grouping. At COP3 held in Kyoto in December 1997, AOSIS countries were first to propose a draft text, in which they called for a 20 per cent reduction in CO_2 emissions on 1990 levels, to be implemented by 2005. The Kyoto Protocol agrees different GHG reduction figures for different Annex I (developed countries, essentially those belonging to the OECD) countries. These range from an 8 per cent cut for the European Union to a 10 per cent increase for Iceland. Combining all Annex I countries targets gives 'a total cut of at least 5 per cent from 1990 levels in the period 2008-2012' (www.unfccc.de/reso...process/components/response/respkp.html November 2001). The European Union can distribute its 8 per cent cut across the various countries as it sees fit. Even prior to the Kyoto Protocol's ratification the UNFCCC website stated that the Protocol's targets were legally binding. However, this was not the case until the Protocol came into force and only then for those countries that have ratified it. It was 31 May 2002 before the European Community and the EU Member States ratified the Protocol and at that time no other Annex I country had ratified the Protocol. Nevertheless, the aim of many people was for the Kyoto Protocol to be ratified and become legally binding, for the right of people to live in a world that is not devastated by climate change to be recognised and for these rights to be fulfilled by legally binding responsibilities. This reflects the ideas of Henry Shue and Jürgen Habermas with regard to the consensual outcomes of discussions being enshrined in rights. The Kyoto Protocol is the result of discursive processes that have involved both pragmatic considerations and ethical constructions.

At the end of COP3 many operational details had yet to be worked out. At COP4 in November 1998 the Buenos Aires Action Plan was adopted with the aim of strengthening the implementation of the UNFCCC and to prepare for ratification of the Kyoto Protocol.

Much of the dispute regarding the Protocol is over the acceptability or not of various sinks (for example forests and oceans are said to be sinks for carbon dioxide). Emissions trading is another bone of contention, as is the idea of clean development mechanisms (CDMs). Emissions trading is when developed countries buy the right of a less developed country to pollute; CDMs are when developed countries or developed world companies, invest in environment friendly technology in less developed countries. The vulnerabilities of less developed countries, both economic vulnerability of some of the developing countries which may be worsened by implementation of the Protocol and environmental vulnerability of the AOSIS and other low lying countries, was also supposed to be addressed after Kyoto. It was agreed that these issues should be addressed and agreement reached before COP6 in November 2000. COP6 was held at The Hague, in the Netherlands; agreement was not reached, blame for this has popularly been placed with the USA, which is notoriously reluctant to forego what is perceived as a right to industrialised lifestyles.

COP6 Part II was held in Bonn in July 2001. The European Union, Germany included, were instrumental in pushing for agreement to be reached to enable the Kyoto Protocol to be ratified. In an interview with a member of the WBGU,[11] the opinion was expressed that it was Jürgen Trittin (Germany's Environment Minister) who really influenced agreement at COP6 II, that he persuaded the rest of the EU Member States' representatives to compromise and gain agreement. This outcome was seen as especially important as the USA under George W Bush's Presidency had declared the Kyoto Protocol "fatally flawed" and had withdrawn from it in March 2001. To enable agreement to be reached concessions had to be made, but it was seen as better that than to effectively destroy the Protocol. The Bonn Agreement included recognition of the need for: new and additional funding (although no specific amounts or legally binding obligation to provide such funds were agreed); and greater flexibility regarding sinks and clean development mechanisms. The Marrakesh Accord was the result of COP7 in November 2001. This basically finalised agreements made in Bonn. There follows a quote from the Marrakesh Ministerial Declaration that is particularly pertinent to this study

> *Believing* that addressing the many challenges of climate change will make a contribution to achieving sustainable development,
> *Recognizing* that the World Summit on Sustainable Development provides an important opportunity for addressing the linkages between climate change and sustainable development.
> .....
> *Request* the President of the Conference of the Parties at its seventh session and the Executive Secretary of the United Nations Framework Convention on Climate Change to continue to participate actively in the preparatory process for the World Summit itself, and to report thereon to the Conference of the Parties at its eighth session. (The Marrakesh Ministerial Declaration, 2001:204-205)

The above quote highlights the linkages between climate change, sustainable development, and therefore, the WSSD and COP8, and therefore, confirms the relevance of the structure of this study.

To enable the Kyoto Protocol to enter into force 55 parties to the convention making up at least 55 per cent of 1990 emissions of Annex 1 parties (industrialised countries) had to ratify the Protocol. Without the USA, which is the largest GHG emitter, ratification by at least the EU, Japan and Russia had to occur. In December 2001 the German government adopted the bill on the enactment of the Kyoto Protocol, with the aim to finish the parliamentary process by May 2002 and to ratify by June 2002. It was the fervent aim of Germany and the EU for the Kyoto Protocol to be ratified before the WSSD. As previously mentioned Germany and the rest of the EU ratified the Protocol on 31 May 2002, Japan followed shortly after on 4 June 2002. It was not until 18 November 2004 that Russia ratified the Protocol, thus paving the way for it to come into force ninety days thereafter on 16 February 2005.

Under the terms of the UNFCCC, each country submits a report on their progress, which is then reported on by an external team. A report on the in-depth review of the national communication of Germany was issued on 21 July 1997 and another on the second national communication was issued on 24 August 1999.[12] These reports are several pages long and cannot therefore be detailed here, however an example of the type of information contained therein follows:

> The amendment of the Thermal Insulation Ordinance came into force on 1 January 1995. The amended ordinance improves the

> energy performance of new buildings on average by 30 per cent (less for large buildings, more for small buildings). This ordinance puts German insulation standards for new buildings among the upper range of European standards. The effects of the amendment to the Thermal Insulation Ordinance were calculated at 8 million tonnes of CO_2 emission reduction per annum (out of a total of 117 million tonnes of CO_2 emissions for all households). (UNFCCC, 1997:12)

With this type of monitoring occurring by an international team, it is likely that countries will want to be favourably reported upon; this presumably encourages environmentally progressive policies to be more favourably considered. Obviously a country will not introduce a policy purely to please the UNFCCC, but any additional encouragement can only be of help, especially if the country concerned is aiming to be seen as a leader in the field as is the case with Germany.

Agenda 21

A brief overview of relevant aspects of Agenda 21 and its formation will now be examined. Prior to UNCED four preparatory meetings were held at which much of the text for the various UNCED agreements was decided upon. PrepCom I was primarily an organisational meeting at which two working groups were established to work on various issues, examples of which are the protection of the atmosphere; of land resources; of oceans, sea and coastal areas; and of freshwater resources. A plenary debated cross-sectoral issues including financial assistance and technology transfer; living conditions of the poor; and education.

PrepCom II built on the work of PrepCom I; it also established Working Group III to cover legal and institutional mechanisms. The plenary spent a great deal of time debating the availability of financial resources and the prospects for technology transfers; the USA, France and Great Britain said it would be unlikely that extra funds would be forthcoming for a general environmental programme. Britain did, however, support the idea of a funding mechanism to help developing countries with regard to environmental problems. (The Global Environment Facility was subsequently established as a pilot programme in 1991; it received additional funding and was restructured in 1994. As previously stated the GEF was proposed by France and Germany.) Progress on Agenda 21 was minimal at PrepCom II.

One of the criteria for Working Group I was the protection of the atmosphere; at PrepCom III there was some unease that too much attention was being placed on energy when transportation, industry, forests and oceans also needed to be considered. One of the proposals to conserve energy was for an energy tax, this the USA opposed saying that 'the discussions focused too much on energy consumption and that the proposals were too interventionist and a potential violation of national sovereignty' (Halpern, 1992:3). There is clearly a justice/liberty debate to be found in this statement, i.e. can global justice be discounted, in terms of it being unjust to cause the degradation and possible decimation of environmentally vulnerable States, in order to preserve the liberty of economically powerful States to continue doing as they will. It is also surely the case that if States are allowed to continue to degrade the environment as they choose, they will in turn obstruct the liberty and sovereignty of environmentally vulnerable States. This latter point may in some cases be literal, as in time some States may actually have no inhabitable land on which to enjoy liberty or sovereignty. At the centre of the realist State-centric rhetoric is that States are sovereign and therefore non-interference in other States is generally said to be a requirement in the international State system. However, the actions of the USA in continuing to be the major contributor of GHGs and thus exacerbating climate change constitutes profound interference in States such as the low lying Pacific Islands. It can be argued that ethical discourse as propounded by Habermas is needed to overcome such limited viewpoints. It will be seen later in this study that Germany actively promotes multi-participatory discourse.

PrepCom IV was a mammoth session that enabled much work to be done to establish agreement on as many issues as possible prior to UNCED. Approximately eighty five per cent of Agenda 21 was agreed. Items not agreed to, included finance and the chapter on atmosphere.

Agenda 21 was written to contain a basis for action, an agenda for action and instruments for action. Its purpose was to be

> the basis for a new global partnership for sustainable development and environmental protection in an increasingly interdependent world. This global partnership must extend beyond the traditional notion of "foreign aid", which is no longer an adequate basis for relationships between rich and poor countries. A whole new relationship is called for, one based on common interests, mutual needs, and common but differentiated responsibilities, one in which developing countries will have both the incentive and the means to cooperate fully in protecting the global environment while meeting their needs and aspirations for economic growth. For this global partnership to be effective, it must be accompanied by new levels of cooperation between all key sectors of society and government. (UNCED, 1992:6)

It is worth noting here that *Deutsche Gesellschaft für Technische Zusammenarbeit* (GTZ – German Society for Technical Cooperation) is a German government owned corporation for international cooperation. The GTZ was established in December 1974. Its aim is to

> Shape political, economic, ecological and social development in partner countries to improve people's living conditions and prospects. (Liptow, 2002)

Whilst this does not denote influence on Germany's part in the inclusion of this objective in Agenda 21, it does show that there was a compatibility of positions. It is extremely likely that Germany would have supported this aim.

Returning to Agenda 21, it was also intended that

> countries whose actions harmed the global environment bore primary responsibility for redressing the damage. While States retain their sovereign right to develop their own resources pursuant to their own policies, their activities must not environmentally damage other States or areas beyond their jurisdiction. States must be held responsible for environmental damage to others from sources within their jurisdiction or control. (UNCED, 1992:6)

This statement is compatible with the polluter-pays principle, which as has already been stated was included in Germany's first Environment Programme in 1971. Again corresponding ideals can be identified, which indicates, that at the very least Germany would have supported inclusion of these ideals in Agenda 21.

Agreement on the sentiments within the above statements was found, but whilst these are noble ideals to work towards actually putting them in to effect is extremely difficult. Once practical considerations come in to play a multiplicity of self serving interests abound and as yet no globally agreed solution has been found and implemented. It has been noted previously that various parties appear to have different views with regard to justice. The wording of Agenda 21 reproduced above would indicate otherwise. Perhaps it is the case that in principle just ideals were agreed upon but that when it has come to putting these ideals into practice, various parties have halted implementation due to self-interest and have called on different interpretations of justice to defend their inactivity. To explain this within a Habermasian framework: consensus was achieved on ideals to be sought, but consensus has not been reached on the instruments needed to fulfil these ideals. Lafferty & Meadowcroft (1996) point out that governments in modern democracies face the electorate every four or five years and that this produces a lack of accountability in terms of the environment. This clearly has detrimental implications for intergenerational justice. Two years before UNCED, Hampson made the following insightful point, which seems to aptly describe the climate change management process.

> Climate change presents new opportunities for international governance and co-operation. An international convention or law of the atmosphere will eventually be required to help govern and care for the global commons, but international co-ordination will be slow and difficult and more modest or 'partial' efforts at international co-operation should not be eschewed in the meantime. (Hampson, 1990:74)

Section 4 of Agenda 21 is entitled Efficient Resource Use; sub-section 4.3 is entitled Energy Resources. This section points out that most of the world's energy presently comes from fossil fuels which are major contributors to GHGs, and that the current course of energy production and consumption patterns will be unsustainable in future years. It is therefore, important that transition to more environmentally friendly energy production be enacted worldwide. Increasing the percentage of clean renewable energy supplies is imperative. It will be seen throughout this study that Germany is active in promoting renewable energy usage. The currently predominant energy suppliers are noticeably reluctant to accept such a scenario. However, Agenda 21 advocates the use of

> a mixture of economic instruments that will provide incentives for energy suppliers and consumers to make environmentally sound choices. From an industrial standpoint, methodologies and criteria must be developed to incorporate the environmental costs of industrial production, including the cost of treating and disposing of generated wastes, into the price of the final products, and incorporate these in the decision-making process. (UNCED, 1992:57)

These appear to be sound suggestions, but again, the realities of implementing such policies are difficult. Nevertheless, as has been previously shown, a number of policies that fit within the above criteria have more recently been introduced in Germany.

Essential means are identified in order to implement Agenda 21. These are: New and additional financial resources; Scientific cooperation and technology transfer; International economy and related domestic policies;

National capacity building; Integrating environment and development decision-making; Strengthening the role of major groups; International institutional arrangements and regional organisations; International legal instruments and mechanisms; and Information for decision-making. It will be seen throughout this study that these issues continue to arise, particularly at international meetings. Discourse was required in order to determine the essential means, it is also apparent that for these to be implemented continuing discourse is required. It can be argued that if a framework is established that facilitates ethical discourse all can come together to work toward improving the circumstances of all. It can also be argued that in such a scenario participatory justice would be fulfilled and a more just outcome in terms of inter-societal and inter-generational justice would result than would be the case in a classical realist situation where the most powerful players win 'hands down'.

The UNCED was a gathering of numerous people including State representatives, scientists, and NGO participants; these people came from a wide variety of backgrounds and had multifarious interests. These people did however, attend a conference on environment and development and to this extent they had a common interest. Within the UNCED process various groups of like-minded people met. A complex multi-layered process occurred, which as argued by Willetts (1996) does not fit neatly into the hierarchical realist model of international relations. Willetts states that 'Individual people, NGOs of a rich variety of sizes and types, governments, the United Nations and other intergovernmental organisations are all interacting in a dynamic manner' (Willetts, 1996:80). He goes on to argue that political values evolve and that 'a theory of social and political change that encompasses all types of political actors and recognizes the interactions occurring between all levels of analysis' (Ibid) is needed. It can be argued that Habermas' discourse ethics meets these requirements as well as indicating how interactions can be improved.

It would appear that globalisation, of amongst other things the dissemination of information and communication technologies, is enabling many more people to participate in transnational politics than has previously been the case. The inclusion of NGOs, business interests and non-State representatives in general, being able to participate in international political processes is variously seen as furthering the democratic process and of examples of a democratic deficit. Mason argues that

> At the global level, NGOs are advancing the democratisation of environmental governance. They are clearly opening up deliberation on environmental norms, transposing domestic

> techniques of public mobilisation across national borders. (Mason, 1999:218)

Mason also states that where NGOs have consultative status at international meetings they are able to circulate information and to make sure that the relevant institutions are accountable to the public. Mason references Bichsel with regard to the point just made. However, Bichsel raises the question of 'legitimacy, accountability and representation' (Bichsel, 1996:234) regarding the claim of NGOs representing popular opinion. She also points out that 'NGOs must receive funding from outside sources, leading to the question of cooptation' (Ibid: 235). Whether or not the inclusion of non-State representatives in international negotiations is interpreted as the widening of democracy, the inclusion of many groups in international regimes can be seen as demonstrative of a more inclusive form of discourse.

The Rio Declaration

The Rio Declaration on Environment and Development includes the stipulation of 27 principles, most of which can be argued to relate to this thesis. However, only those that have a close fit with the overall themes of justice will be reproduced here. The relevance of these principles will be evident throughout this study; they still permeate international negotiations and whilst not all countries adhere to them, Germany, appears at least, to endeavour to implement them.

- Principle 2 – States have, in accordance with the Charter of the United Nations and the principles of international law, the sovereign right to exploit their own resources pursuant to their own environmental and developmental policies, and the responsibility to ensure that activities within their jurisdiction or control do not cause damage to the environment of other States or of areas beyond the limits of national jurisdiction.
- Principle 3 – The right to development must be fulfilled so as to equitably meet developmental and environmental needs of present and future generations.
- Principle 4 – In order to achieve sustainable development, environmental protection shall constitute an integral part of the development process and cannot be considered in isolation from it.
- Principle 15 – In order to protect the environment, the precautionary approach shall be widely applied by States according to their capabilities. Where there are threats of serious

or irreversible damage, lack of full scientific certainty shall not be used as a reason for postponing cost-effective measures to prevent environmental degradation.

- Principle 16 – National authorities should endeavour to promote the internalization of environmental costs and the use of economic instruments, taking into account the approach that the polluter should, in principle, bear the cost of pollution, with due regard to the public interest and without distorting international trade and investment. (UNGA, 1992)

The continuing UNCED process

Following the UNCED, the Commission for Sustainable Development (CSD) was formed in December 1992. Its remit was to monitor and report on the implementation of the UNCED agreements at local, national and international levels. It was also agreed that a review of the progress since UNCED would be conducted during a special session of the UNGA in 1997. This special session (Rio+5) took place in June 1997 at the UN headquarters in New York. This will not be analysed in depth, suffice to say that the Resolution adopted by the General Assembly included the statement that the world's environment had continued to deteriorate since the Rio conference and that

> Significant environmental problems remain deeply embedded in the socio-economic fabric of countries in all regions. Some progress has been made in terms of institutional development, international consensus-building, public participation and private sector actions and, as a result, a number of countries have succeeded in curbing pollution and slowing the rate of resource degradation. Overall, however, trends are worsening. (UNGA, 1997)

The resolution also made the point that the special session was intent on accelerating the implementation of Agenda 21.

During Rio+5 Chancellor Kohl put forward the suggestion that a world environment organisation be established. This idea is one that is still in vogue and will be discussed in greater detail in the next chapter, with regard to aims for the World Summit on Sustainable Development (WSSD or Rio+10).

The conference in Rio was called the United Nations Conference on Environment and Development, although that particular conference finished on 14 June 1992 the UNCED process continues. Ten years after the Rio Conference, the WSSD was held in Johannesburg and the eighth Conference of the Parties to the UNFCCC was held in New Delhi. The following chapter examines the formulation of Germany's aims in the lead up to these conferences.

4

PRE-WORLD SUMMIT ON SUSTAINABLE DEVELOPMENT (WSSD)

Forming of Germany's climate change related aims for the WSSD

As the previous chapter has shown Germany's preferred course in the evolution of international climate change politics is the product of many years of influences and events. It should also be noted that whilst Germany's position (as that of any other country and the international community as a whole) will have evolved over many years, it is also the case that the international relations of climate change are inevitably dynamic. Thus Germany's position has also to be dynamic and capable of reacting to events over which it has little or no control. The official German position is also the result of a multiplicity of domestic inputs; some of which will be examined, as will the aims of the German government for the WSSD. COP8 was held approximately two months after the WSSD. Whilst the outcomes of WSSD would affect to some degree events at COP8, overall aims were clearly going to be very similar, if not identical. Hence analysis in this chapter that is specifically focused on the WSSD is also relevant to COP8.

It will be seen that discourse ethics are in many ways implemented in environmental politics in Germany, and that justice considerations, at least implicitly, are drawn upon. The information given in both the last chapter and in this one demonstrates that German environmental policies are fairly well advanced, thus giving Germany credibility with regard to its aims for the international arena.

The first section of this chapter will examine various domestic inputs to the climate change debate that are deemed to be possible sources of the German position. The positions of research institutions, whose work is brought to the notice of the relevant sections of government, will be assessed. Appraisal of the positions of non-governmental organisations that work on climate change will follow. Business and industry are affected by climate change related policies and this sector of society will be considered next. The final sub-section prior to the summing up of this section will analyse information from government sources.

Research Institutions

As noted in chapter three the German Advisory Council on Global Change (WBGU) was established in 1992, it provides the government with academic research and advice and presents relevant government departments with the reports it produces. Official meetings are held at which Ministers receive the reports and comment upon them. It has been confirmed that WBGU reports (particularly the first paper to be considered in this section) were considered by those people working within the government on the German position for the WSSD.[1] The stance taken by the WBGU will, therefore, be the first to be considered within this section. In the autumn of 2001 the WBGU produced a policy paper entitled 'The Johannesburg Opportunity: Key Elements of a Negotiation Strategy'. Items within this policy paper that relate to climate change issues will be detailed. The introduction states that the responsibility for global climate change must be

> primarily attributed to the world's energy systems. Climate protection policies are doomed to failure unless countries rethink and restructure their energy systems, including transport. For this reason, a key component of global sustainability policies should be to develop a global energy strategy aimed at restructuring global energy systems; (WBGU, 2001:3)

Attributing responsibility implies that there is a duty of care. World energy systems per se clearly cannot be held to be responsible, but the people that control them, and arguably the people that use them can. The German government is making some progress in directing the supply of energy to come from renewable resources. The government is also intent on all sections of society becoming involved in the pursuit of efficient use of energy. Such matters will be dealt with towards the end of this chapter. It will also be seen later in this study that Germany was particularly active at the WSSD with regard to the promotion of renewable energy.

The paper also notes that issues such as climate change, scarcity and pollution of freshwater supplies are interconnected and that these issues will further impact upon health, for example the spread of infectious diseases in certain areas, which will in turn impact upon the economies of the affected areas. Networked policymaking structures are identified as being necessary to deal with such problems. The report states that

> Existing institutions are too fragmented and under-financed to meet these challenges effectively. The WBGU therefore believes that the priority goal for the World Summit on Sustainable Development must be to reform the organizational and financial basis of international environmental and development policy. (Ibid)

As will be seen throughout this chapter a number people and groups share this goal. It is also a part of the official governmental stance.

> Feasibility concerns dictate that priorities be set regarding the specific content of such policymaking. The Council therefore recommends that the WSSD strive to adopt selected milestone agreements on particularly important issues, ... The WBGU appeals for the following priorities to be set:
> 1. *organizational reform* of the international environmental policy system,
> 2. mechanisms for adequate and reliable *funding* of international environmental policy,
> 3. developing a *global energy strategy* aimed at intensifying climate protection efforts,
> 4. strengthening a world-wide operating system of protected areas and advancing the development of a *Protocol on Forests,*
> 5. adopting a *global freshwater strategy,* and
> 6. convening a world summit to address the implications of global environmental changes for infectious diseases. (WBGU,2001:3-4)

Whilst all of these priorities do not directly deal with climate change, they are all connected with it. More specific recommendations that are pertinent to this study follow.

With regard to reforming the international environmental policy system it is stated that the programmes of the Global Environmental Facility (GEF), which is the funding body for international environmental agreements and is administered jointly by the World Bank, the United Nations Development Programme (UNDP) and the United Nations Environment Programme (UNEP) are inadequate. Organisational and financial reforms are called for, along with improved scientific support; these reforms should be coordinated. It is worth noting that Germany likes to work with and through the GEF, this will be expanded on later in this study, when it will be seen that Germany has, along with Canada, instigated reorganisation of GEF's monitoring and evolution unit. The report states that UNEP should be reformed so that it is autonomous, though still associated with the UN. An organisation along the same lines as the World Trade Organisation (WTO) is envisaged, to be called something like Global Environmental Organisation or World Environmental Organisation (WEO). This is a policy that Germany pursues, however, as stated in chapter three, Chancellor Kohl suggested in 1997 at Rio+5 that a WEO be established. Thus, the inclusion of this in the WBGU's 2001 document can at most be seen as support for continuing this policy aim. The WBGU report states that a WEO should give equal representation to member parties as it is only by giving effective representation to developed and developing countries that acceptance and therefore, effectiveness of a WEO would be achieved. It is also pointed out that all environmental meetings could be held where the WEO's secretariat was based, this would enable poor countries to attend more meetings. At present it is beyond the means of some countries to attend conferences at various locations around the world. If meetings were always held in the one place, it would be feasible for many countries to situate a team of specialists permanently in that place. This would in effect widen democracy, as more people would be able to participate in the discourse involved in agenda setting and negotiating. This would also correspond with Habermas' contention that all parties should be able to participate in discourse, thus advancing the potential of consensus being agreed upon, and decisions made that all parties could accept. It may be the case that consensus is easier to achieve if a limited number of people are in discussions, but those left out of the process are less likely to agree with the outcomes, because they would not have had their concerns listened to and thereby, addressed.

The report details recommended steering and financing mechanisms for implementation at international level. These include charges to be made for the use of the global commons; these charges should be used to reduce environmental damage and to pay for adaptations needed due to environmental impact caused by such use. Later in this chapter it will be seen that in 2002 a Ministry for Economic Cooperation and Development (BMZ) report stated that the German government favours researching the possibility of the introduction of charges for the use of global environmental goods and for these charges to be a part of a new financial mechanism.

The plight of developing countries is expanded upon by pointing out that it is they that suffer most from global change. Developed nations are called upon to substantially increase the financial support given to developing countries as was agreed at UNCED in 1992. The paper calls upon Germany to increase its development cooperation funding. It will be seen later in this chapter that Germany has made a commitment to increase overseas development assistance. It will also be seen that an International Conference on Finance and Development was held in March 2002 at which this issue was discussed. It is probable that a number of influences contributed to the official German position and that the WBGU's advice contributed to the policy-making discourse.

To enable the achievement of sustainable development requires environmental protection and the world economic order to be coterminous. The paper holds that a liberal world economic order could promote sustainability by adopting measures that

> internalize environmental externalities at the national and global level. … this necessitates international procedures for ensuring the reconcilability of multilateral environmental treaties and WTO rules. However, it must be stipulated in clear, unmistakable terms that environmental protection has priority over free trade principles when the two conflict.
>
> The precautionary principle must be assigned greater weight in the WTO rules than is currently the case, (WBGU, 2001:6-7)

To this end it is advocated that the matter of reconciling WTO regulations with global environmental goals be placed on the WSSD agenda. It is envisaged that a strong WEO should act as a counterweight to the WTO. It will be seen that these aims were reflected in Germany's goals for the WSSD.

Agreeing on a global energy strategy is the subject of section three of the WBGU policy paper, the first sub-heading of which is 'Protect climate successfully by transforming energy systems.' This is a necessity, as it is the burning of fossil fuels that is the largest contributor of GHGs and hence to global warming. The current predominance of producing energy by burning fossil fuels must therefore, be addressed. As already mentioned Germany proactively promoted the expansion of renewable energy at the WSSD.

The report also states that

> The distribution of reduction commitments in the future should be based on clearly defined and acceptable criteria, such as 'equal per capita emissions' as a basic equity principle. In detail, consideration could be given to different energy consumption needs, depending on climate zone, potentially misguided incentives for population growth, and the risk of excessive quotas. (WBGU, 2001:8)

Analysis of COP8 will show that Germany advocated entering into discussions on future commitments on an equitable basis. Whilst the notion of 'equal per capita emissions' would appear to be fair; it would be inordinately difficult, if not impossible to either agree upon or in fact to implement. The last sentence of the above quote points to the need for complicated formulae through which acceptable and viable quotas could be reached; to some extent the Kyoto Protocol can be seen as the forerunner for such a scenario. It should be borne in mind that there are differing viewpoints as to what is acceptable climate change. At different ends of the viewpoint continuum can be found: the government of the USA, which is willing to accept current levels and trends in climate change (this could be argued against as in many ways they just deny that anthropogenically caused climate change is happening); and the low-lying small island States that are already finding climate change unacceptable, some of these islands are already suffering salt water inundation in ground water, others are already losing land to the seas. It can be argued that, for there even to be a possibility of agreement on an acceptable level of climate change, the ideals of Habermas' *communicative rationality* need to be employed in negotiations.

The WBGU paper calls upon the German government to work towards an international convention that agrees upon the phasing out of

> energy price subsidies and subsidies on non-sustainable energy technologies (oil, coal, nuclear energy and large-scale environmentally unsustainable hydropower). The deadlines for adaptation must be specified in a binding manner, taking into account the socio-economic and ecologic situation in the respective countries at a relevant baseline. (WBGU, 2001:9)

Targets for renewable energies are advocated, as is the promotion of the efficient use of energy. It is argued that incentives are essential to create renewable energy markets and to restructure the energy industry, but that these should be in place for a limited time and should reduce over time. Karsten Sach[2] of the BMU also holds this view; Sach made the point that subsidies should gradually decrease so that they have a built in incentive to be successful. The paper calls for incentives to encourage research and development into new renewable and efficient energies. It is also pointed out that long-term continuity regarding renewable energy markets and energy efficiency is necessary, toward this end long-term targets should be established. With regards to policies through which to attain targets it is suggested that national and regional strategies

> be developed, disclosed and discussed within a network for exchanging experience. A special role is played here by energy suppliers. As global players with major financial resources, they

> must be integrated as partners, and agreements concluded to provide a framework that is binding on their operations. (WBGU, 2001:9)

Again, it can be pointed out that in order to progress towards such eventualities widespread multi-participatory discourse is required.

The invigoration of climate protection by modifying transport policies is the next item in the policy paper. It is pointed out that thus far the role of transport has been neglected in policymaking (presumably the focus here is on international policy making, as Germany has introduced some policies with the aim of reducing the increase in vehicular transport). It is argued that projected growth rates in the transport sector threaten the fulfilment of commitments made by the developed States under the Kyoto Protocol. The paper calls for the WSSD to

> adopt principles for sustainable development in the transport sector. The Federal Government of Germany should appeal for a Programme of Action on Transport policy, in which the objectives for sustainable transport systems are also specified.
> ...Developing countries aiming to expand their transport infrastructure should receive guidance on how to avoid unsustainable developments and on using sustainable leapfrogging technologies. In order to speed up the pace of international negotiations, the Federal Government of Germany should exert pressure so that the European Union imposes a levy on kerosene fuel in aviation prior to the WSSD, even if other states are not immediately in favour. This could fortify Europe's leadership role on the environmental issues. (WBGU, 2001:10)

In 2002 Germany confirmed support of the debate for a tax on aviation fuel to be introduced. The word 'should' is often used in WBGU literature; this implies moral duty/responsibility, which in turn implies that if such responsibilities are not fulfilled that injustices will be committed.

The summary to this paper includes

> The Federal Government of Germany has repeatedly acknowledged its global responsibilities in the field of environment and development policy. In the run-up to Johannesburg, the Federal Government can accept the challenge, as part of the European Union, of a leading role in the global sustainability debate and in that way act commensurately with the increased importance of our country in international politics since 1990.

> There is much more to such a decision than the ethical and humanitarian extension of German foreign policy. Given the burgeoning of interrelationships and mutual interdependencies in the world, it is in Germany's immediate interest that the risks inherent in global change be warded off by preventive and curative strategies and programmes implemented by the international community. The Federal Government cannot achieve a high level of security, sustained prosperity and social stability unless it is prepared to cooperate closely on policymaking at the international level.
>
> Strategies for surmounting the problems that exist have a price. The politicians, business community and population in Germany must accept the fact that international efforts will require higher levels of funding than has hitherto been the case. (WBGU, 2001:17)

The language used by the WBGU can again be seen to be emotive; however also cited are foreseen practical necessities for needing to take action. This type of dialectic can also be seen in much of the output from government sources. Speeches made by both Chancellor Schröder and the Environment Minister, Jürgen Trittin in relation to the WSSD are prime examples and will be addressed later in this study.

Similarities between the advice in the report and government policy and action have been indicated in the preceding text. It would appear that the WBGU policy paper 'The Johannesburg Opportunity: Key Elements of a Negotiation Strategy' has had an impact on the government position. It should, however, be remembered that this paper was just a part of the discourse that was in existence in the lead up to the WSSD and even though the WBGU is an advisory council to the German government and correlations between the advice and the actions exist, this does not prove a causal link.

In 1999 the WBGU produced a Special Report entitled *Welt im Wandel: Umwelt und Ethik* (World in Transition: Environment and Ethics). On 17 December 1999 the report was handed to the Research Minister, Edelgard Bulmahn of the SPD and to the Environment Minister, Jürgen Trittin of the Green Party. The WBGU argues that all people and cultures value quality of life, although different perceptions of what is required for such may be held. Likewise views on what constitutes environmental risks, it is argued, differ between countries. As views within countries on such matters can be bitterly contested, it is presumed that the WBGU report defines countries in this instance as the official State apparatus. In the report it is argued that although differences in perceptions may be held, the fact is that quality of life is important to all, and that recent research shows that basic principles regarding this matter are remarkably coherent across cultures. Thus it is argued, the pursuit of universal principles for the protection of the global environment is reasonable. The German government is entreated to support this idea. It is pointed out that the global replication of German environmental policies is not an aim but that basic principles similar to those in the Declaration of Human Rights should be aspired to.

> Conflicts in weighing up a particular situation occur frequently. Therefore the WBGU differentiates between *categorical* principles, which must not be violated under any circumstances, and *compensatory* principles, for which a balance with other competing principles may be acceptable. (WBGU, 1999:2)

The language used in the above paragraph is reminiscent of Immanuel Kant's categorical imperative. A categorical imperative is one that is necessary in itself, irrespective of any end product; it is something which one would will to be a universal law. The use of the term 'categorical principles' by the WBGU could indicate that terminology used by Kant has entered into more widespread discourse. As indicated in the introduction, Habermas recognises the importance of linguistic and communicative abilities for effective discourse and it can be argued that the use of language that calls for ethical implementation is perhaps a first step toward actual ethical implementation.

The development of ethical expertise was instigated by the government. Such issues are clearly important to the government, and it can therefore, be surmised that the above report would have been carefully considered.

The work of the Heinrich Böll Foundation will now be examined. Although included in the research institution section the Heinrich Böll Foundation is not purely a research institution though it does provide support for research. The foundation is a legally 'independent political foundation' (it is affiliated to the Green Party) its

> primary objective is to support political education both within Germany and abroad, thus promoting democratic involvement, socio-political activism, and cross-cultural understanding. (Heinrich Böll Foundation, 2001:34)

As noted above the foundation is affiliated to the Green Party, during the WSSD preparatory process the Green Party were part of the coalition government and as already stated, a member of the Green Party, Jürgen Trittin, was Environment Minister; it is likely, therefore, that during the formation of the government position, consideration would have been given to papers produced by the foundation.

In the run up to the WSSD the Heinrich Böll Foundation published a series of 'World Summit Papers'. It has been confirmed that these papers were considered within the BMZ in the lead up to the WSSD.[3] Not all of these are relevant to this thesis, but pertinent items from selected papers will be looked at. 'From Rio to Johannesburg: Contributions to the Globalization of Sustainability' paper 5 actually contains three separate papers, two of which will be considered in some depth. It was mentioned at the beginning of this chapter that analysis of the inputs to the German government stance would be undertaken. Both Trittin and Eid were representatives of the German government and are attributed authors of the two papers to be examined. The inclusion of analysis of these papers in this section along with inputs from non-governmental sources could, therefore, be questioned. However, whilst these pieces are presumably representative at least to some degree, of government opinion, it is also the case that this opinion will be constitutive of these individuals' viewpoints as well as other inputs, it is also the case that both Trittin's and Eid's paper's were written, at least partially, in the first person thus declaring personal ownership of these views and not claiming governmentally held beliefs. Whilst these papers may well be mainly reflective of the government stance, because they are personal they can be, and sometimes are, more radical than the official government position. It is extremely likely that these personal views would also be made within governmental settings and thus would be a part of the discursive policy-making process. Another point to be made is that by publishing quasi-governmental viewpoints, opinion will be given on these and it is therefore, possible that it could be further refined. By placing in the public domain such opinions, these can be engaged with and commented upon and possibly developed; such a process would be demonstrative of discourse ethics in action.

The first paper to be addressed here is by the Environment Minister, Jürgen Trittin, and is titled 'Towards a Global Sustainable Energy Policy'. As if in reply to the often heard arguments that environmental protection does or will harm economic development it is stated that

> A destroyed environment threatens sustainable economic development and social justice. Protection of the environment and the natural resources is therefore a prerequisite for a development that is viable for the future. (Trittin, 2001:22)

Implicit in the above paragraph are issues of intergenerational justice,

It is argued that the UN structures for environment and sustainable development need to be reformed as a matter of urgency. Trittin's personal opinion is that it is vital to strengthen the UNEP and thus the consideration of the environment within the UN system, and that to this end, the idea of transforming the UNEP into a World Environment Organisation should be evaluated. The strengthening or transformation of the UNEP must be accompanied by formalised committed financing; it should not be, as is currently the case, reliant on voluntary contributions. As has already been stated the creation of a WEO is advocated by the German government.

Dealing specifically with climate change, an assessment is given of the current situation during which it is pointed out that the use of renewable energy resources globally is a tiny proportion of the total energy used; it is stated that

> We share the view of some major energy producers like Shell and BP/AMOCO that it is possible to raise the share of these environmentally sound energy sources to fifty per cent within the next fifty to sixty years. (Trittin, 2001:23)

It is worth noting that the term 'we' is used, presumably Trittin is voicing a governmental opinion (as well as his own) at this point. The paper immediately goes on to argue that it is his opinion that nuclear power is very risky and that he does not believe it is necessary to use such power to supply energy to either industrialised or developing countries. Storage of waste products from nuclear energy production 'places a heavy burden on future generations' (Ibid). Again ethical values concerning intergenerational justice are apparent. It should be noted that Germany is phasing out the domestic use of nuclear energy. Later in this chapter it is argued that the exclusion of nuclear energy from the Kyoto Protocol's Clean Development Mechanism was due to pressure from Germany.

It is argued that a new energy approach is needed, which includes increasing energy efficiency, saving energy and increasing renewable energy usage. Germany's efforts at the WSSD will be seen to be reflective of this statement. Action that Germany is taking to change from reliance on fossil fuel and nuclear energy towards renewable energies is then discussed. The national climate protection program that was instituted in October 2000 is outlined, as are the promising results of this programme to date.[4] This piece of writing shows pride in the achievements thus far, it is stated that the target reduction of 25 per cent of CO_2 emissions on 1990 levels by 2002 are on track, and that this is being achieved by real reductions not by the manipulation of figures. Nevertheless, it is acknowledged that the measures so far undertaken are insufficient to fully address the issue of climate change and the needs of sustainable development. It is argued that efforts must continue beyond the 2012 targets, and that toward this end further investment is needed along with changed patterns of consumption and production. The importance of public support is acknowledged especially as emissions from private households and from transport are increasing. It is stated that Germany is trying to lead the way with regards to climate protection, but that global action is required to address these issues, and that binding international laws are required to ensure such action. It is acknowledged that Germany cannot solve global environmental problems alone, and that if other countries fail to take action when Germany is active, then Germany would suffer in terms of competitiveness. The Kyoto Protocol is not named at this point, but the fact that calls for ambitious commitments to be accepted post 2012[5] alludes to the protocol. It is then stated that

> We expect our partners in the industrialized world to acknowledge our joint responsibility. (Trittin, 2001: 26)

This reference to joint responsibility is consistent with the polluter pays principle. It can also be interpreted as the adoption of a moral stance, especially as concern is then expressed over the USA's position regarding the Kyoto Protocol. The paper concludes

> Those who ignore Kyoto will miss the chance for positive developments with regards to economic growth and employment.
>
> Our experience in Germany shows clearly: Climate protection creates jobs. According to a very conservative estimate by an economic institute, up to 200.000 jobs can be created in the period 2000 to 2020.
>
> I expect from the World Summit a strong impetus for a complete restructuring of industry and society. There is a need for broad participation of the whole society. NGOs have an important role to play in defining requirements for a sustainable development. Therefore, I call on all NGOs to participate in the preparations for Johannesburg. (Trittin, 2001:26)

It is worth noting that this paper resulted from an NGO forum, it would, therefore, have been geared toward an NGO audience, which may have been a reason to emphasise the role of NGOs. Nevertheless, the call for multi-participation is reflective of discourse ethics, and Germany's consensual style of politics.

The paper by Uschi Eid is entitled 'Key Issues On The Road to Earth Summit 2002'. Eid was Parliamentary State Secretary to the Ministry for

Economic Cooperation and Development; and a member of the Green Party. The paper calls for

> a strong and clear commitment to the goals and principles of Rio on the part of industrialized countries. Current patterns of consumption, as we all know, are profoundly unequal and unjust, so the industrialized countries have to be prepared for considerable changes in their way of life and production, and they have to agree to concrete steps in this direction. (Eid, 2001:28)

Calls for action with regard to righting injustices are made in this statement. Effectively Eid is calling for industrialised nations to take responsibility for their actions; this can be related to Shue's argument, as explained in the introduction, that duties exist to avoid depriving; to protect from deprivation; and to aid the deprived. It could be argued that this is rhetoric and that the required actions do not necessarily follow, but it can also be argued that ideals need to be held and rhetoric engaged in for discourse ethics, as advocated by Habermas, to have a chance of being accepted and implemented. Eid's assertions largely mirror Trittin's. The paper calls for industrialised countries to be subject to binding emissions limits; for adaptation to climate change to be effected especially in poor regions; and for action to be taken toward sustainably fulfilling the developing world's need for energy. Eid, with her 'governmental hat' on, points out that the BMZ is already contributing some DM 200 million a year to the latter, and that support for energy efficiency, including energy sector reforms is provided. Germany's rhetoric in international negotiations reflects Eid's argument. Efforts that Germany makes to assist climate change adaptation in poor regions and the support it gives to energy reforms in the developing world will be examined in chapter seven.

The Heinrich Böll Foundation papers that are examined above contain a mix of praising government policies and calls for advances in policies. It is these latter qualities that can be considered as inputs to the climate change debate and that the government may well have considered.

Non-governmental organisations

There are many NGOs; clearly it would be impossible to investigate and analyse the positions and actions of all of them. However, the work of a selection of the main organisations will be looked at. The reason for each NGO's selection will be given when they are introduced.

To start with, relevant papers published by the *Stiftung Entwicklung und Frieden* (SEF - Development and Peace Foundation) will be examined. The SEF was founded in 1986 by Willy Brandt (SPD member and Federal Chancellor between 1974 and 1982); its working principles are: global responsibility, an interdisciplinary perspective and cross-party dialogue. The working principles of the SEF are pertinent to the themes that run through this study, and therefore, contribute to the decision to examine SEF outputs, as do the following factors. As the SEF was founded by Willy Brandt and the SPD were in power in the run up to the WSSD, it is a distinct possibility that the works of the SEF would have been a part of the discourse considered during the policy-making process. It is also the case that SEF

> projects are targeted to decision-makers and opinion leaders in politics, academia and the business community, as well as to civil society multipliers and key media figures.
> (www.sef-bonn.org.en/about/profile/index.php?print=3 October 2004)

The SEF sends all its publications to the executive and to government ministries, mainly to the relevant unit but also to high-ranking officials in the Foreign Office, the BMZ and where appropriate to the BMU and any other appropriate ministry. Publications are also sent to MPs, *Länder* ministries, European policy-makers and to international organisations.[6] It has been confirmed that the papers examined in this study were considered by the BMZ.[7] It has been stated that the papers may have been considered by the BMU but it was noted that as the papers examined in this study were published some time prior to the WSSD preparation process, they may not have been very influential in that process.[8] It can, however, be argued that the ideas within these papers would have been already been considered, possibly taken on board and thus be a part of the prevailing discourse at the time of the preparation process.

Although not directly written for the WSSD, the second policy paper that the SEF published in 1996 does have some relevance to this subject matter. Its title is 'Global Governance: Challenges to German Politics on the Threshold of the Twenty-First Century'. A few points are worth noting, one of which is the recognition that international regimes such as the UNFCCC are important elements in global governance. In this paper international regimes are defined thus:

> international regimes, in which states enter into contractual agreements to tackle shared problems. (Messner & Nuscheler, 1996:7)

It is also recognised that if problem-solving requires global efforts

> Then this can only be done on the basis of specific, universally recognized values and principles. …
>
> The joint solution of problems in network-like systems of negotiation presupposes a minimum of trust both within national societies and within the international system. (Messner & Nuscheler, 1996:8)

It is surely the case that to reach a situation of trust and universally recognised basic values and principles, discourse is required that, as far as possible, meets with the criteria as detailed by Habermas. It can be seen that the discourse ethics, (which are essentially about justice of participation) and consensual politics that are aspired to within Germany are relevant to the efficient workings of international relations.

It is argued that

> Because of the increasingly bitter competition on the world markets, there is a danger that the great global-governance projects in development and the environment will be abandoned … (Messner & Nuscheler, 1996:9)

This may seem like pure comment on the state of world systems, and in some ways it is; in fact the comment goes on to state that nationalistic policies are prevalent and points to the anti-cooperativeness of this state of affairs. It can also be seen as prophetic with regard to the stance taken by the USA, particularly under the George W Bush Presidency, in climate change negotiations. However, the relevance to this section of this study is that it gives the background as to why there is a call both later in this paper and in others, for the German government to call for UNEP to be developed into a Global Environmental Organisation. As has already been stated Germany advocates the creation of a WEO.

It should also be noted that Nuscheler (co-author of the paper) as a member of the WBGU was involved in the WSSD preparation process.[9]

'A World Environment and Development Organization' is the title of the SEFs policy paper 9, published in June 1998. Here it is argued that such an organisation should combine UNEP and CSD and the secretariats of relevant international conventions. This, it is argued would give the new organisation greater status, greater capacity for the negotiation of, and implementation of agreements, and greater financial efficiency. As well as the financial savings that the integration of various organisations and secretariats would allow, various additional funding opportunities are identified. An example of which is the idea that levies be imposed on international air travel and international financial transactions and that these should be automatically directed to this new organisation. This paper was published in 1998, as already mentioned similar policies are advocated by the WBGU in a report dated 2001. Chancellor Kohl suggested creating a WEO 1997, and the German government have more recently supported the idea of taxing aviation fuel. Calls by various parties for the introduction of similar policies make it difficult to apportion definite influence to any one party when the policy is adopted. It is likely that increasing evidence of favoured policies in the prevailing discourse provides cumulative weight to the argument for the policy. It should be noted that although the policy paper is from the SEF, its authors are not those of the previously discussed paper. The authors of this paper are Frank Biermann and Udo Ernst Simonis.

It is pertinent to give some background information on the authors of the SEF papers examined above. Dr Dirk Messner is Executive Director of the Institute for Development and Peace, at the Gerhard Mercator University, Duisburg; Professor Franz Nuscheler is Director of the same institute and as already mentioned a member of the WBGU. Dr Frank Biermann was at the time of publication based at the Belfer Center for Science and International Affairs at the Kennedy School of Government at Harvard University, though noted as a member of the secretariat of the WBGU. In 2003 he became Professor of political science and environmental policy sciences and Head of the Department of Environmental Policy Analysis at the Institute for Environmental Studies at the Vrije University in Amsterdam. He also directs the Global Governance Project, a multidisciplinary international research programme of the Potsdam Institute for Climate Impact Research, the Free University, Oldenburg University and the Vrije University. Professor Dr Udo Ernst Simonis is Research Professor for environmental policy at the Social Science Research Centre in Berlin. The point of detailing the various positions of these authors is to demonstrate that a variety of academic opinion and expertise is harnessed. It can therefore, be argued to be indicative of discourse ethics in action as the ideas of a variety of authors are published and promoted by the SEF.

Policy paper 6 of the SEF is by Dr Thomas Fues, who is a research fellow in sustainable development at the Institute for Development and Peace. The paper's title is 'Rio Plus 10: The German Contribution to a Global Strategy for Sustainable Development.' Fues argues that economic globalisation is detrimental to the Rio process. It is stated that

> The 'jobs or sustainability' alternative, for example, is a spurious one. Anyone who wants to be properly equipped to face the future has to bring the two objectives into harmony with one another. The transition to a path of sustainable development can and must be politically configured in such a way that a society's ecological

> and social innovativeness and its economic competitivity used are simultaneously strengthened. (Fues, 1997:2)

The second half of the last sentence could be used as a definition for sustainable development. Whether or not the position taken by Trittin and the German government has its origins in this piece of work, the sentiments of it are reflected in Trittin's speeches.

Fues makes the point that although agreed courses of action have to occur at the global level, it is recognised that each State should decide on their own methods of implementation in line with their own capabilities, priorities and specialisms. This viewpoint is also expressed in government documentation.

With regard to Germany, Fues calls upon the Federal Government to implement policies at the national, European and global levels that can make a substantial contribution to the Rio process, and to encourage the implementation of such policies at regional and local government levels. The Federal Government is called upon to work towards the introduction of an EU wide energy/CO_2 tax and for the EU to impose a kerosene tax on international air travel. At national level the government is called upon to carry out

> Ecological reform of energy legislation, in order to increase the scope for demand-management, renewable energy sources, and power-heat link-ups, and in order to bolster the independence of local energy supplies. Realignment of public support-measures, in order to bring about radical increase in energy productivity. Sustainable mobility through the maintenance and expansion of the national rail-network. (Fues. 1997:8)

Regional and local governments are also called on to introduce various policies to use and promote renewable energy, public transport usage, more environmentally friendly ways of distributing goods and the following of ecologically sound standards for land-use and development. The policies advocated have since been evident in German policy-making, as can be seen from policy description in the previous chapter and later in this chapter. The implementation of ecologically sound policies in Germany is argued to be crucial with regard to the influence that the Federal Government can exercise at the international level. The argument that Germany's domestic policies act as an example and give credibility to its aims in international relations can be found in speeches given by Trittin.

The paper calls on the government to

> work out a national sustainability strategy on the basis of Agenda 21. This draft plan will contain precise goals, timetables, and implementation scenarios that will serve as a basis for a participative social discourse. The cornerstone of the scheme will be a national environmental plan. (Fues, 1997:10-11)

Other measures such as: the setting up of committees of sustainable development within the Bundestag, regional parliaments and local councils; industry requiring to prove its "readiness to help shape the ecological and social regulatory framework for sustainable development" (Fues, 1997:11); and the need for NGOs and other civil groups to make the most of their resources in mobilising public support are detailed. As will be seen later in this chapter, a national sustainability strategy has since been drawn up and accepted; its development included 'participative social discourse'. A Council for Sustainable Development was established during the formation of the national sustainability strategy.

Many of the policies advocated for the Federal Government, have been adopted by them. Whilst no direct causal link can be established between this paper and government policies, there is a correlation between the two and it is possible that this paper was at least a part of the aggregate information that led to the adoption of such policies. As already mentioned SEFs 'projects are targeted to decision-makers and opinion leaders in politics' (www.sef-bonn.org.en/about/profile/index.php?print= 3 October 2004). This aim, the correlations of advice and policy, and the similarity of rhetoric, point to potential linkages between this policy paper and the government position.

The work of another NGO, or more precisely a forum of NGOs will now be looked at. The *Forum Umwelt und Entwicklung* is the German NGO Forum on Environment and Development. It was established by thirty five organisations in December 1992, six months after UNCED in Rio. Its founding objectives were:

- Taking seriously the outcome of Rio and to try to do whatever possible to eradicate poverty world-wide and to protect the environment;
- Lobbying both at national and international level to implement the decisions passed in Rio, particularly Agenda 21;
- Establishing working groups which, for example, develop position papers on the most pressing issues in the Rio follow-up;
- Coordinating education and information programs;

> • Increasing pressure on government and legislative bodies by joint NGO actions;
>
> • Acting as a contact for international partners.
>
> The plenary of the NGO Forum on Environment and Development meets annually to prepare its further activities. The Secretariat is in charge of coordination and information. (www.forumue.de/aboutus/secretariat/index.html October 2002)

Clearly this organisation is significant as through it the 'voices' of many smaller NGOs are channelled, thus giving them a greater combined voice. The Forum's founding objectives, particularly those of developing position papers, lobbying at national level and increasing pressure on government, make analysis of any impact its work may have on the German government's position relevant to this study. The secretariat of the Forum is partially funded by the BMU and the BMZ: two points come to mind, firstly that perhaps the independence of that voice may be compromised;[10] the second point is that as the BMU and BMZ commit funds to the Forum, they are likely to take note of its activities and outputs. It has been confirmed that the BMU regularly receives a circular from the Forum and that the BMZ takes note of the Forum's campaigns.[11] It was also pointed out that Jürgen Maier, who is Director of the Forum's secretariat, and/or his colleagues were members of the German delegation at CSD meetings and at WSSD PrepComs. This would appear to be yet another example of discourse ethics in action; it would seem that such ethics have been implemented in an institutionalised manner. The Forum on Environment and Development has, for the above reasons, been identified as being a useful organisation to monitor for this research, specifically the working group on climate.

> The working group sees itself as a focal point of exchange for German NGOs that focus on national, European and international climate policy. It bundles up the work of the NGO, in order to make it better known to a broad section of the public, as well as to the political and social decision makers. (www.forum.de/topicandworkgroups/climatechange/index.html November 02)

The Forum has produced a great deal of information, and it is possible only to mention some items and selectively expand upon others.

In the lead up to WSSD the Forum on Environment and Development placed an article entitled 'Shaping global equity on environmental terms: 10 Points for Sustainable Development' on the website www.rio-10.de. This report announces the intention to campaign for 'Global equity on environmental terms'. It is also stated that the German government is expected to promote the points raised at national, European and global levels. Climate change related issues are led by the second point, which is 'Protecting the world's climate by redirecting energy and transport policies.' Industrialised countries are called upon to ratify the Kyoto Protocol and for this to come into force before the WSSD. It is argued that action by industrialised countries should enable a 40 per cent reduction in GHGs by 2020 and an 80 per cent reduction by 2050. The report argues that the proportion of energy sourced from renewable resources should increase both nationally and globally by in excess of 10 per cent within the next ten years. As will be seen when analysing COP8 in chapter six, Trittin give a speech in which he stated that Germany would commit to a 40 per cent reduction in GHGs by 2020 as long as the EU reduced emissions by 30 per cent and other countries adopted similar targets. Trittin also said that Germany was committed to increasing renewable energy production to 12.5 per cent by 2010. Whilst the words Trittin used were not identical to those used in this document, the positions are very similar.

Debate within the international arena is advocated regarding taxes on international travel, the idea being that revenues from such taxes should be used for environmentally and socially sustainable development. As already stated this argument has been made by other parties, it appears to be an element of much of the environmental discourse in Germany. As already stated the German government advocates research into the possibility of introducing such a scheme.

'Strengthening Local Agenda 21 processes and participation at all levels' is the subject of point ten. This is largely a point directed at German domestic policy and as such will not be dealt with in depth. An issue worth lingering on is that of

> The right of access to information and social participation in environmental policy have been stipulated in binding form under the international law for the first time in the Europe-wide Aarhus Convention, which entered into force in 2001. We call upon the German government to ratify the Aarhus Convention swiftly. In Johannesburg, a process should be initiated to establish the provision of this convention at the global level, too.
>
> Elements of direct democracy need to be strengthened or introduced. Procedures for true public participation must be understood as a communicative process and not as a formal act. (www.rio-10.de/rioprozess/texte/10_points_for_johannesburg.html August 2002)

The last sentence is essentially calling for multi-participation in ongoing discursive processes and is therefore, calling for major components of discourse ethics to be implemented. It can be argued that this sentence indicates that discourse ethics as an ideal is embedded in German thinking, at least with regards to academic and political circles that are involved in the pursuit of environmental sustainability.

The Aarhus Convention is the common name for the Convention on Access to Information, Public Participation in Decision-Making and Access to Justice in Environmental Matters;[12] it was opened for signature on 25 June 1998 in Aarhus, Denmark. It entered into force on 30 October 2001; Germany signed on 21 December 1998 but as of December 2005 has yet to ratify it. This could be seen as indicative of a gap between government rhetoric and implementation. Nevertheless, in a speech in July 2001, Minister of State Volmer appeared to provide German governmental backing for the Aarhus Convention by stating that

> Crisis prevention is not solely a task for governments. Cooperation between the public and private sectors plays an important role. The 1998 Aarhus Convention is intended to strengthen the role of civil society in decision-making on environmental issues. A sound environmental policy is part of this development. Against the background of our own experiences in Germany I can say with pride that the activities of the ecological movement in the eighties and nineties played a crucial part in ensuring that today environmental protection has become an indispensable element in the policies of every German government. (Volmer, 03-07-01)

This excerpt also confirms that German governments have taken note of concerns that citizen participation in environmental issues brought to their attention. In other words, multi-participation in environmental discourse has been effective.

As part of a public awareness-raising campaign for the WSSD, and also as a way of making government aware of, at least a section of public opinion, a campaign of action was held in Berlin on 31 May–2 June 2002.[13] The motto for this action, which itself was described as a central element of the campaign *'Globale Gerechtigkeit ökologisch gestalten'* (To shape global ecological justice), was *'Zukunft für alle - Mitmachen bewegen, verändern'* (Future for all – to take part in, to move, to change). The action campaign included:

- On 31 May *Unser Klima geht baden – wer haftet?* (Our climate goes bathing – who is responsible?), this is described as *Ein Kaleidoskop zum Start der Klima-Ausbade-Kampagne von Germanwatch* (a kaleidoscope of who carries the can regarding the climate campaign). Germanwatch is an NGO.

- On 31 May–1 June a 'Countdown for Johannesburg' conference was held; this was organised jointly by The Forum on Environment and Development and the Heinrich Böll Foundation. The themes of the conference were energy, nourishment, globalisation and mobility. The conference was billed as enabling participants coming from different perspectives to find common ground on standards of German politics for the WSSD – for a future for all.
- On 2 June a bike ride was held to raise awareness of the campaign's political demands for the WSSD.
- Also held on 2 June was an environment festival.

Various merchandising was available for the campaign, which could also be purchased by post.

A climate caravan was conducted as part of the campaign. It started in Berlin on 31 May 2002 (at the activity campaign mentioned above) and finished in Bremen on 17 October 2002, having travelled through various German towns. Banners that were used included 'our climate goes bathing – who is responsible?' and 'the South should not carry the can alone for climate change'. The caravan was clearly an awareness raising exercise; it was also

> looking for answers to the questions "Who is liable for the damage and willing to shoulder the costs of adaptation measures?" and "How could a structure of political and financial accountability look like?" The climate caravan is an opportunity for discussion with victims from North and South as well as a chance for dialogue with insurers and those responsible for climate change. (Germanwatch, 2002)

Germanwatch credits *Evangelischer Entwicklungsdienst* (Protestant development service) and *Die Nordrhein-Westfälische Stiftung für Umwelt und Entwicklung* (North Rhine Westphalia Foundation for Environment and Development) with promotion of the climate caravan campaign, and the BMZ with support. It is also worth noting that in an interview with Christoph Bals[14] of Germanwatch it was stated that Germanwatch works closely with the people in the German government, for example, within the BMU and the BMZ, that are working on climate change. It was stated that contact occurred between Germanwatch and a senior climate change negotiator within the BMU approximately on a weekly basis. This is another indicator that multi-participative discourse is conducted in the formation of Germany's official position with regard to the international relations of climate change.

It is clearly the case that by the time the above campaign was staged, the German government would have arrived at its aims for the WSSD. As previously mentioned, policy would have been evolving over time. Presumably these last minute activities were aimed at: raising awareness; consolidating government opinion; and ensuring that government were aware that much of their domestic constituency wanted them to be proactive at the summit.

An activity that the Forum engaged in where there is proof that the German government took note, is the sending of a letter on 5 August 2002 to Chancellor Schröder, the contents of which were given in a press release on 6 August. The letter discusses the then forthcoming WSSD and the progress made since UNCED. Here it is stated that developing countries post Rio had hope, but since then this has been transformed into disappointment, and that a failure in Johannesburg would send out a disastrous signal. It is also stated that at UNCED Germany announced GHG reductions of 25 per cent by 2005 and that this had been a powerful sign that led to an international dynamic which led to the Kyoto Protocol. It goes on to state that the time has come for comparable signs at the WSSD and that concrete targets should be called for. The letter implores: Germany to increase its share of renewable energy in its power consumption from 6 per cent to 25 per cent by 2015; to increase its development aid from 0.27 per cent of GNP to 0.7 per cent by 2010; and to reduce its emissions of GHGs by 40 per cent of 1990 levels by 2020. The letter appeals for action and a successful summit in Johannesburg. The letter was signed by the following organisations: *Arbeitsgemeinschaft der Umweltbeauftragen der Ev. Kirche in Deutschland, AGU; Arbeitsgemeinschaft Regenwald und Artenschutz, ARA; Evangelischer Entwicklungsdienst, EED; Forum Umwelt und Entwickling; Freunde des ASA-Programms; Germanwatch; Misereor; Ökumenische Initiative Eine Welt; Pro Regenwald; terre des homes Deutschland; Umweltstiftung WWF-Deutschland; Verkehrsclub Deutschland, VCD; Verband Entwicklungpolitik Deutscher Nichtregierungsorganisationen, VENRO; Weltwirtschaft, Ökologie und Entwicklung, WEED; Weltfriedensdienst; Weltladen-Dachverband.*[15]

A reply was sent from Joachim Nick-Leptin of the Chancellor's office, to Jürgen Maier of the Forum on Environment and Development on 13 August 2002.[16] In the reply it is stated that the Federal Chancellor had during the past weeks and months emphasised many times that the WSSD offered a great chance to shape the welfare of humankind. It is also stated that within the preparatory process framework the German government belonged to the group of States that stressed the need to make the Johannesburg conference a success. Together with European partners, Germany would push for concrete targets and an action programme to: increase the share of renewable energies; halve the number of people without access to sanitation; institute an action programme for sustainable water and the energy industry; to institute a ten year action plan for sustainable consumption and production. The letter goes on to mention the government's strategy for sustainable development, details of which are dealt with later. This letter is evidence that the German government considered the contents of the letter sent by the aforementioned NGOs.

Business/Industry

It appears there were no orchestrated business campaigns, at least none that were in the public domain, prior to, and regarding the then forthcoming WSSD. The lack of a campaign could be indicative of a perceived lack of importance of conferences, which are heavy in rhetoric and do not necessarily result in the implementation of policies. It is likely that business interests will be dealt with through ongoing relationships between various businesses, business groups and government ministries. It could be assumed that the absence of visible pressure could denote that power with regard to agenda setting exists. Lukes (1974) argues that agenda setting is a potent form of power, all the more so because it is not transparent. However, Joachim Hein[17] of the *Bundesverband der Deutschen Industrie* (BDI - Federal Association for German Industry) posited that there were many ideological, good and eager people in the BMU, but that they only took note of the environment and that they should consider broader issues such as jobs and the economy. This viewpoint is indicative that traditional business interests do not have a great deal of unseen influence with the BMU, which is the lead ministry with regards to the position taken on climate change at the international level. It is worth noting that the positions taken by different sectors of industry vary considerably. Traditional industries, such as coal mining will be largely against the sorts of policies advocated by those people who are pushing for climate change mitigation, i.e. the reduction and eventual elimination of subsidies for non-renewable sources of energy. Obviously, the renewable energy industry will be in favour of progressive climate change related policies. It is also the case that, at least certain sections of the finance and insurance industries are in favour of policies that aim to mitigate climate change; examples of which are Munich Reinsurance and Dresdner Bank. It is pertinent to note that both Munich Reinsurance and Dresdner Bank have representatives who are members of the UNEP Financial Initiatives Climate Change Working Group.

Government sources

Before examining information gained from government sources it is worth reiterating that the official government position will have been arrived at through a cumulative long-term process. The evolution of policies as detailed in the previous chapter, together with more contemporary inputs as described in this chapter will have contributed to Germany's policies.

The BMU is the lead ministry with regards to climate change issues, although the Foreign Office (AA), the Ministry for Finance (BMF), the BMWA and the BMZ are consulted. Papers are distributed to each of these ministries and any details that are disagreed with are discussed and a consensus agreed upon. The BMU coordinates this process. Generally the BMU is keen on progressive policies, the BMF focuses on the financial implications of proposals, whilst the BMWA watches the language used regarding commitments to reduce emissions. Mentioned in chapter three was the fact that Beuermann and Jäger (1996) hold that industrial influence with various ministries together with inter-ministerial differences weakens the environmental policies of the BMU. It does appear to be the case that some industries have more favourable relations with different ministries; for instance it is likely that the renewable energy industry and the BMU have a good working relationship, whilst traditional industries such as the coal industry appear to have a better relationship with the BMWA which is more concerned with the economic health of the country and maintaining good industrial working relationships. However, as indicated in the previous section, the BDI does not believe that the aims and policies of the BMU are significantly changed by the interests of mainstream industry. Although different ministries have different focuses, climate change is an issue that all agree needs to be dealt with, and so any differences tend to be regarding the degree to which commitments are made. The actual details of this process regarding the specific documentation for the WSSD and COP8 have not been ascertained.

Observable governmental positions regarding the WSSD will now be examined.

On 8 December 2001 the Environment Minster, Jürgen Trittin, spoke at the Berlin Conference on the Human Dimensions of Global Environmental Change, which was specifically addressing 'Global Environmental Change and the Nation State'. The speech he gave was not addressing the then forthcoming WSSD, but some pertinent points were made, including the assertion that national policies can have considerable influence with regard to global environmental change. Trittin used as an example the case of persistent organic pollutants (POPs); in 1972 Germany was the first country to ban the use of a POP (DDT - dichloro-diphenyl-trichloro-ethane), which was subsequently followed by the banning of other POPs. Following this the European Commission brought in directives banning POPs and eventually UNEP under Klaus Töpfer initiated a convention to phase out the production of twelve POPs, this was adopted in Stockholm in 2001. Although citing this as a success story, Trittin recognised that such a lengthy process is not affordable for each environmental concern.

Trittin stated that Germany works actively within the EU and that at global level it is the Council Presidency and Commission of the EU whom engage in negotiations. Trittin argued that the united front presented by the EU enabled the continuation of the Kyoto process at Bonn, despite the USA trying to block the process. He also pointed out that

> Communitarisation does not exclude taking a frontrunner role in environmental policy. (Trittin, 2001)

He proceeded to give examples of environmental policy that has been implemented or decided upon within Germany, such as the decision to phase out nuclear power and the promotion of renewably sourced energy. He argued that these policies and others that make climate protection policy successful within Germany were the

> Reason that we were successful in enforcing our wish in the negotiations in Bonn that the construction of nuclear power plants in countries in the South will not be recognised as an offsetting measure within the Clean Development Mechanism. (Trittin, 2001)

This corroborates the assertion at the beginning of this chapter that Germany's domestic policies give credibility to its aims in the international arena.

Speaking about the WSSD, Trittin talked of

> initiating a policy geared towards a decentralised energy supply and renewable energies. It may even be possible for us to agree on different yet feasible annual quotas for each country for increasing the share of renewable energies. That sounds a lot easier than it is. Whenever the national budget or well-organised major groups are concerned, there is a potential cause for conflict. (Trittin, 2001)

This comment can be seen as prophetic, as will be seen when the WSSD is being discussed.

Trittin went on to state the need for strengthening international environmental organisations and environmental law, and of the need for these not to be overpowered by the decisions of the WTO. German support for the transformation of UNEP into a WEO was stated. It was argued that the funding for a WEO should not be purely sourced from contributions from Member States, but also from mechanisms employing polluter pays principles with regard to the use of valuable global goods, by for example aircraft and maritime transport. This is reflective of arguments put forward by the WBGU and Biermann and Simonis.

Trittin argued that in a world where global companies are more powerful than nation states

> The historic task of nation states today is therefore to introduce global environmental legislation that is more powerful than any nation state or any transnational corporation. (Trittin, 2001)

He stated that perhaps the main problem with regard to protecting the world is the identification of most people with the nation-state, and the voting in of governments on narrow national policies, not on vital international issues such as climate change. The development of global awareness was argued to be imperative. It could be argued that raising global awareness would lead to widespread discourse being engaged in, this could then lead to consensus on the need for, and type of action required, and ultimately for action to be taken. If this were to occur it would in many ways reflect the previously described events that occurred in Germany following the widespread realisation of *Waldsterben.*

On 19 December 2001 Hans Martin Bury, Minister of State to the Chancellor, presented a draft national sustainability strategy entitled 'Prospects for Germany'. Bury is reported as having said

> that sustainability is "the main theme of reform policy for this government"… "Intergenerational justice, social solidarity, quality of life, and international responsibility are the coordinates of our strategy". (German Press and Information Office, 2001: Press Release No. 588/01.)

It is clear that issues of justice and responsibility are important for Germany and that ethical considerations are not limited by national boundaries or to any large degree, by time.

As the final version of the strategy will be discussed below, details of the draft report will not be examined.

During the formation of the national sustainability strategy, systems were in place to enable multiple viewpoints to be considered. An Enqeutte Commission 'Sustainable Energy Supplies in View of Globalization and Liberalization' was established; so to was the Council for Sustainable Development, which was made up of seventeen members from different sectors of society. In addition to these consultation bodies and talks with major interest groups, an on-line forum to enable dialogue was set up by the Council for Sustainable Development. This avenue for public input was mentioned at the presentation and in the related press release.

Advertising this opportunity enabled the possibility of greater public participation in the policy-making process. It is, therefore, demonstrative of consensual politics in action and augurs well for the effectiveness of the policy as it has received widespread input and agreement. This is an example of the importance of society's ideas, being considered through participative dialogue in the formation of policy that is promoted on the international stage. This example is also indicative of the implementation of discourse ethics as put forward by Habermas.

On 17 April 2002 the Federal Cabinet approved the national sustainability strategy, which was to form the German contribution to the WSSD. An article that was placed on the government website states that

> The impetus for the federal government's sustainability strategy came from the 1992 UN Conference on Environment and Development in Rio de Janeiro, at which the international community of nations accepted the principle of sustainable development and established a global action programme for the 21st century, Agenda 21. This calls upon the signatory states to develop a national sustainability strategy with the goal of achieving an economically powerful, socially equitable, and ecologically sound form of development.
> (Federal Government, July 2002.
> http://eng.bundesregierung.de/documente/Artikel/ix_7632.htm
> November 2002)

The article goes on to state that

> In the coalition agreement of 1998 the federal government had already taken up this United Nations assignment and made clear that it wished to formulate a national strategy for sustainable development in dialogue with the most important groups within society. (Ibid)

The reflexivity of discourse between various levels, i.e. local, national and international, is evident.

'Perspectives for Germany: Our Strategy for Sustainable Development' is the title of the final version of the national sustainability strategy, relevant aspects of which will now be discussed. In the foreword it is stated that sustainable development is a way of directing globalisation to enable social, ecological and economic well-being, and that by presenting its national strategy at the WSSD, Germany is making an important contribution to this aim. The strategy is detailed, setting out aims, priorities and measures to be taken to achieve such aims. Clearly within the confines of this study, it is only feasible to outline certain points therein. It is stated that

> This strategy will be the basis for further political reforms and also for changes in the way businesses and consumers behave. Extending far beyond the ecological challenge, this strategy will also serve to provide guidelines for a comprehensive policy, able to meet the challenges of the future, to fulfil responsibility, which bridges across the generations, for devising a form of development which is economically, ecologically and socially sustainable.
> (www.nachhaltigkeitsrat.de/service/download_e/pdf/Perspectives_for_Germany_short_version_pdf p3. December 2002)

Evidently, ethical matters are seen as being important. The very notion of sustainable development implies notions of intergenerational justice.

> The ethical basis of sustainability is that every generation must deal with its own problems and should not burden the coming generations with them.
> (www.nachhaltigkeitsrat.de/service/download_e/pdf/Perspectives_for_Germany_short_version_pdf p5. December 2002)

Sustainability requires the participation of all sectors of, and members of society, each person and institution needs to consider the results of their actions. Here as well as the call for mass participation, a reminder is given that along with citizenship, which in itself confers the right or even duty to participate, goes correspondingly a certain amount of responsibility. Echoes of Shue's (1980) proposition that for every right there are corresponding duties are evident.

The strategy includes rules for managing sustainability. Required areas of action that are identified and are pertinent to this study include the following. In the long term, consideration will be given to the regenerability of all renewable natural resources that are used, and when using all non-renewable natural resources such as fossil fuels, consideration will be given to how these functions can be fulfilled by the use of an alternative. It is also foreseen that in the long term, the products of such resource usage will not reach a level that is detrimental to the world's ecosystems. The decoupling of energy and resource usage from economic growth is seen as a necessity, as is the offsetting of the growing demand for energy, resources and transportation by increasing efficiency.

It is envisaged that

> The international framework conditions are to be devised in such a way that people in all countries can lead a life of human dignity according to their own cultural principles and can participate in economic developments. Environment and development form a unity. In an integrated approach the fight against poverty is to be linked
> - to respect human rights
> - to economic development, environmental protection, and also
> - to responsible action on the part of Government.
>
> (www.nachhaltigkeitsrat.de/service/download_e/pdf/Perspectives_for_Germany_short_version_pdf p10. December 2002)

The section titled 'Use energy efficiently – protect the climate effectively: Scenario for an energy policy sustainable in the future' gives figures relating to energy consumption, thus explaining the need for reform. It is stated that Germany together with the EU have via international agreements undertaken to significantly reduce GHG emissions. It is reported that Germany is well on its way to meeting its 21 per cent reduction target by 2008/2012, as in 2000 emissions were 18 per cent down on those of 1990. It is also acknowledged that the national goal is to reduce these emissions by 25 per cent by 2005 and that the current reductions are not yet close enough. Increased efficiency in production and consumption of energy is aimed for. Details of how efficiency of production is to be achieved with regard to modernisation of power plants and promotion of innovative technology are detailed. Also detailed are the increases in efficiency of motor vehicles and domestic appliances. A 'concrete goal' to double energy productivity by 2020 from 1990 levels is laid down.

'Development of renewable energy sources' is the title of the next section. It is stated that since 1998 wind power capacity has trebled; that the future of wind energy is offshore; and that demand for solar panels is high. It is the government's policy to encourage development of renewable energy. Government targets include: the doubling of the proportion of renewably sourced energy in primary energy consumption by 2010 from 2000 levels; and for the percentage of renewably sourced energy to reach approximately 50 per cent by the middle of the century.

The facts that nuclear energy is highly risky and waste products cause problems for generations are stated and it is therefore, affirmed that nuclear energy is not a viable alternative to conventional forms of energy production. The phasing out of nuclear energy production within Germany is assured.

The problem of increasing vehicular transport is outlined, as is the need for the reversal of this trend. A cause of increased goods traffic is identified, as is the manner in which production has been organised with increasing divisions of labour so that many suppliers are involved in the production of an item such as a refrigerator. It is envisaged that greater movement of goods between west and east will be an outcome of the forthcoming expansion of the EU. Efficiency of transportation thus needs to be increased, together with the decoupling of economic growth from increasing transport levels. A target has been set for goods transport to reduce by 5 per cent by 2020 based on 1999 levels. This target is recognised as being difficult to achieve as goods traffic has been increasing. Toward this end transport costs must include the stress caused by such transport on roadways and the environment. Strategies used to fulfil this aim have already begun, for example, mineral oil taxes have been gradually rising; in addition, it is stated that tolls are to be introduced for heavy lorries on selected autobahns. A doubling of goods to be transported by rail by 2015 from 1997 levels has been set as a target. This would mean that 25 per cent of all goods transported would be by rail. Since 1998 there has been increased investment in the railways, with an extra three billion euros being invested between 2001 and 2003. The prospect of rail and vehicular transport being used in combination is being encouraged and so to is the creation of combined transport terminals.

The report makes it clear that sustainability strategies must take into account the effects of domestic actions on other areas of the world and expresses its support for the establishment of a United Nations World Commission for Sustainable Development that would ensure that continued globalisation occurs within the confines of sustainability.

With regard to financing overseas development, it is stated that

> efforts are being made to make available, within the framework of the available budget, at least 0.33 per cent of the gross domestic product for public developmental aid.
> (www.nachhaltigkeitsrat.de/service/download_e/pdf/Perspectives _for_Germany_short_version_pdf p36. December 2002)

Support is expressed for the aim of reaching 0.7 per cent as quickly as possible.

With specific reference to the WSSD it is stated that Germany is committed to developing strategies to increase the worldwide use of renewable energy and to increase energy efficiency, and to

- Strengthening the financial and economic basis for Sustainable Development in developing countries by means of trade facilitation measures and developmental collaboration,
- Joint initiative with the German economy, trades unions and non-governmental organisations to take greater account of social and environmental factors and also of Sustainable Development principles in making direct foreign investments,
- Development of the structures of the United Nations in the area of Sustainable Development with the additional aim of involving developing countries more effectively in the future direction of the processes of globalisation,
- Improved co-ordination of environmental activities in international institutions and within a framework of international agreements,
- Strengthening of the United Nations Environmental Programme (UNEP) with a view to developing it into a world environmental organisation.

(www.nachhaltigkeitsrat.de/service/download_e/pdf/Perspectives_for_Germany_short_version_pdf p37. December 2002)

Progress reports on developments made with regard to the national sustainability strategy are to be made biannually. It is also recognised that a sustainable development strategy must be an ongoing process, with the strategy being tested regularly. New developments, for example in technology or the international arena must be taken into account. The report finishes with the following rallying cry.

> Thus Sustainable Development remains a continuing task for politics and society. The Federal Government is addressing this task and encourages all social groups to take an active part in the sustainability project, and, in the interest of our children and grandchildren, to make their own contribution towards a Germany which is able to meet the challenges of the future in the One World.
> (www.nachhaltigkeitsrat.de/service/download_e/pdf/Perspectives_for_Germany_short_version_pdf p40. December 2002)

This government document is clearly arguing for intergenerational justice to be considered and implemented. It is not only a rallying cry to the German populace but a policy statement and thus shows that the government intends implementation.

In May 2002 Chancellor Schröder whilst speaking at a conference of the Council for Sustainable Development proclaimed that 'Germany is set to fight for more "global justice" [and that] Global justice would become a "question of survival" in the twenty first century.' (BBC via Financial Times, May 2002) Speaking about the WSSD 'Schröder demanded that the Johannesburg conference adopt "an action programme that can be implemented" and give a "starting signal for a sustained energy supply".' (Ibid) It was reported that he said

> It is up to politics to prevent a division into winners and losers of globalization – in Germany and on an international scale, ... That is why Germany supports the initiative to stock up the funds for global environmental protection, which will involve an additional 2.7bn dollars for the developing countries in the coming years. Based on the resolutions of the recent EU summit in Barcelona, the EU will stock up funds for development cooperation by 11bn euro by the year 2006. (BBC via Financial Times, May 2002)

There were no calls from opposition parties for dramatically different objectives to be pursued at the WSSD. This reflects the similarity of positions across the political spectrum with regard to climate change politics. As far as aims are concerned, both the government and the opposition follow the same objectives. Such convergence coincides with Habermasian expectations of discursive political processes. This convergence does not however, pertain to the priority placed on instruments through which to reach objectives. The SPD/Green government favour renewable energy subsidies and have decided to phase out nuclear energy. The opposition CDU do not agree with the support given to the renewable energy industry and are in favour of using nuclear energy. However, the opposition did not feel strongly enough about these issues to campaign on them in their 2002 election campaign. The SPD did include climate change issues in their election campaign.

During 2002 the BMZ placed on their website a report entitled 'From Rio to Johannesburg: A review of progress in selected areas of German development policy'. In this report it is stated that

> German development cooperation is committed to the guiding vision of sustainable development and its translation into tangible action through the agreements reached in Rio. (BMZ, 2002a)

This is another instance that is demonstrative of the reflexivity of ideas and policies between the international level, the national level and indeed foreign policy.

In the section on protecting the environment it is acknowledged that the energy sector is a major source of GHGs and that these cause damage to the climate. It is also noted that globally in excess of two billion people are without secure energy supplies. It is argued that more people must have access to energy and that this increase should be met in a sustainable manner. The report states that Germany can make a substantial contribution to this challenge as it is amongst the world leaders in renewable energy technology innovation. It is stated that Germany provides 102 million euros annually for promoting renewable energy sources in the developing world. Germany is also stated as being the third largest contributor to the Global Environment Facility,[18] which in turn is said to provide US$200 million per annum for climate protection in developing countries. Support for the Kyoto Protocol's Clean Development Mechanism is pronounced, as are the:

> Focal areas of German development cooperation in energy policy:
> - Providing policy advice in the energy sector
> - Promoting renewable sources of energy
> - Improving rural energy efficiency
> - Raising energy efficiency
> - Improving conventional power station technologies (BMZ, 2002a)

The financing of sustainable development and the undertaking of industrialised countries made at Rio to increase their overseas development assistance (ODA) to 0.7 per cent of GNP is discussed. There has been a general trend of declining ODA but it is noted that Germany began to reverse this trend in 1998 and that it will endeavour to contribute 0.33 per cent of GNP by 2006. It is pointed out that Germany is

> the world's third-largest bilateral donor of overseas development assistance (ODA) and has considerable influence in the multilateral development organisations. (BMZ, 2002a)

An International Conference on Financing for Development was held in Monterrey in March 2002, at which it was agreed to reverse the trend of declining ODA; prior to this meeting the EU agreed to raise its ODA to 0.39 per cent by 2006, from its current 0.33 per cent.

This section of the report ends with confirmation that the German government supports the debate regarding the establishment of new financing methods such as researching the idea of a currency transfer tax, and the introduction of charges for use of global environmental goods, i.e. airspace and oceans.

The report goes on to argue that traditional politics focuses on short term outcomes, and the retention of power, and that it is at least partly for this reason that no country has yet implemented a consistent sustainability policy.

> a paradigm shift in political control systems [is needed]. German development cooperation assists partner countries to take this step. (BMZ, 2002a)

Institutions that are able to cooperate and participate in strategic policy management are seen as being a solution.

> German development cooperation has concentrated (initially in the environment sector) on enhancing the capability of institutions, including NGOs and associations, to take necessary action, as well as on supporting cooperation initiatives and long-term policy processes. (BMZ, 2002a)

An example is given of German support for national poverty reduction strategies to include consideration of the environment and improving such to become sustainability strategies. Germany is increasing support to partner countries that are planning to introduce national sustainability strategies.

> The aim is not to formulate additional plans and programmes, but to optimise the planning processes in the partner countries. They have to be designed along strategic, participatory and cross-sectoral lines... (BMZ, 2002a)

The ideal of participatory justice can be detected in this policy aim. This point is more explicitly made in the following section where it is acknowledged that for sustainable development to be successfully implemented, widespread participation at all levels is needed. Developing countries need to be able to participate in: international negotiations; benefits of globalisation; and in the use of global resources. Citizen participation is also seen as a necessary requirement for successful policy implementation.

> The political framework for this cooperation must be founded on the principle of ownership and the process-oriented character of

> participatory development as stressed in the OECD paper "Shaping the 21st century".
>
> German development cooperation incorporates the intensive participation of target groups at an early stage. This applies both to project planning and steering and to the promotion of participatory policy approaches in partner countries, such as national sustainability strategies...(BMZ, 2002a)

Participation is again called for, both for developing countries and for citizens. As has been previously mentioned, participation in negotiations falls within the realm of discourse ethics. Participatory justice is advocated, as is intergenerational justice, albeit it implicitly.

The report points out that to achieve sustainable development, structural change is also required in industrialised countries, Germany included. Education is important and it is stated that

> Equal attention is now being paid to education for both environmental and development policy. Decisive pointers in this direction were set at the congress 'Education 21 – learning for a just and sustainable development', organised jointly by the Association of German Development NGOs (VENRO), the BMZ, the Standing Conference of the Ministers of Education and Cultural Affairs of the German States, and the responsible ministries in the German states. (BMZ, 2002a)

VENRO is a member of the *Forum Umwelt und Entwicklung*, the activities of which have previously been discussed. The above-mentioned congress is an example of the implementation of discourse ethics.

The report also states that

> Between 1999 and 2002 the BMZ boosted funding for its information and education work by 90 per cent to ca. Euro 4 million. (BMZ, 2002a)

It is also pointed out that

> Numerous development projects are being implemented by German NGOs in cooperation with local partner NGOs and with financial assistance form the BMZ. (BMZ, 2002a)

Dialogue between the BMZ and NGOs is ongoing, with the former cooperating with approximately three hundred NGOs each year. Even so, the BMZ holds that institutional structures for cooperation with NGOs need strengthening. It is stated that the BMZ sees the activities of NGOs as being reflective of public opinion; presumably this is one of the reasons why the BMZ deem working with NGOs to be so important.

Conclusion

The primary focus of this chapter has been the domestic formation of Germany's aims in the international relations of climate change. It is important to ascertain how Germany's aims have been formed, as this impacts on how Germany negotiates in the international arena. It is also the case that the formation of these aims is interconnected with domestic policies that have been successful, for example the promotion and development of renewable energy, this imparts credibility to Germany on the international stage and thus adds to the influence that it is able to achieve.

It has been shown that Germany's aims have resulted from an ongoing process which is characterised by its multiple inputs and influences. Whilst all elements of Habermas' discourse ethics have not been fulfilled (for all interested parties to have equal access to decision-making discursive processes and for ideal speech situations to exist, are ideals that would be extremely difficult, if not impossible to implement), discursive inclusiveness has been evident. In other words, participative justice has been shown to be implemented to a large degree in the formation of Germany's international aims.

The positions outlined in this chapter overlap to a large extent; it can be argued that this is due to the participative and discursive nature of the German political system with regard to environmental politics. Obviously not everyone agrees on every point; there are some policies and actions that for example NGOs have advocated that have not been taken up by the German government, or at least not to a great enough degree to satisfy NGOs. A criticism by the World Wide Fund for Nature (WWF) Germany, is that the national strategy is

> "well-meant, to be sure, but too little to really get any thing going". … The granting of loans and guarantees must be subject to stricter criteria on nature conservation and environmental protection. "For example, it is absurd that German money is helping to finance the felling the last rainforests," the WWF says (Guist, 2002:10)

A criticism from Friends of the Earth Germany (BUND) is that targets for reduction of goods traffic and GHG emissions are too low.

Nevertheless, the German government's position does reflect the various inputs as previously outlined. It should be remembered that discourse is undertaken to reach optimal decisions that can be consensually agreed upon. If there were truly an ideal speech situation, complete knowledge was available and there was unlimited cognitive ability then perhaps all could conceivably agree completely. Such a situation has not yet been reached (nor may ever be), which in turn means that it is unlikely that everyone involved in discussions will agree totally with every aspect of outcomes. This chapter has however, shown that discourse ethics have in large part been implemented in environmental negotiations within Germany; examples of which include: reports provided by the WBGU; the support for the work of the *Forum Umwelt und Entwicklung* by the BMU and BMZ; the government reply to the letter dated 5 August 2002 to Chancellor Schröder from numerous NGOs regarding policies for the WSSD; and by the participative social discourse that was implemented during the formation of the national sustainability strategy, and the fact that this was recommended by Fues in a paper published by the SEF in 1997. Many participants have been able to make their opinion known to the policymakers, and hence these opinions have been available for consideration. As has been mentioned throughout this chapter, implementation of discourse ethics, mass participation and justice considerations is occurring. This may only be in infancy and more must be done if full implementation of such ideals is to occur, but such ideals are clearly being considered, and acted upon to some degree.

It has been noted that the German government advocates a strong WEO, one that has power over the WTO and indeed over nation states. It is clear that it is believed to be better for the future of the world that the power of the WTO is tamed and consideration for the environment be enshrined in international law. These issues were advocated by the WBGU, Trittin in a Heinrich Böll Foundation paper, and Biermann and Simonis in an SEF paper. The *Forum Umwelt und Entwicklung* called for environmental and human rights agreements to have primacy over the WTO. Whilst it cannot be proven that these sources influenced the German government, there is a correlation between these positions. It should however, be reiterated that as detailed in chapter three, Chancellor Kohl had suggested the formation of a WEO during Rio+5 (in 1997). The willingness of the German government to cede power to a WEO could be indicative that Germany is less afraid of a 'Federal World' than many other States because it is used to multi-layered governance within its own borders. The term 'Federal World' has been used in the context of legally binding international agreements and laws with regard to events that need global governance. Undoubtedly the German government wishes to retain national sovereignty over national events and strategies.

This chapter has looked at the formulation of aims to be worked towards at the WSSD. COP8 followed less than two months after the WSSD and hence the aims for the former were largely the same as the climate change related aims for the latter. Before examining events, and Germany's influence at both the WSSD and COP8 in chapter six, the following chapter will appraise Germany's relations with the EU.

5

GERMANY AND THE EUROPEAN UNION

The European Union (EU) is not the focus of this study; however, Germany is a member of the EU and as such works within the framework of the EU. At international negotiations such as those assessed within this study, the Member States of the EU pre-agree a unified position that is presented by the EU delegation. It is therefore, necessary to consider relations between Germany and the EU. It should be borne in mind that such relations are not akin to relations with another State as Germany is a constituent of the EU. The first section of this chapter will provide a brief overview of the evolution of EU environmental politics, particularly those relating to climate change. This will be followed by an examination of the development of the EU stance regarding climate change international relations. Germany's role in the policy-making procedure will also be considered. It should be noted that when talking with actors about the EU, the international level is defined as relations between the EU and non-EU States.

Evolution of environmental policy in the EU

In the 1957 Treaty of Rome, which created the European Economic Community (EEC) there was no mention of the environment, and therefore, no specific authority for environmental policies to be introduced. Nevertheless, over a number of years it became apparent that damage to the environment and, therefore, environmental protection needed to be considered. At the Paris Summit in 1972 ministers decided that the Community should take measures to address environmental issues. It is worth noting once again that in 1972 the UN Conference on the Human Environment was held in Stockholm at which time there was much activity in environmental politics. Germany had introduced its first Federal Environment Programme in 1971. Germany was an important member of EEC (and still is of the EU) and as environmental policy-making had taken a major step forwards in its domestic politics, Germany was proactive in encouraging the EEC to take action with regard to the environment. Von Seht and Ott (2000) argue that

> German environmental policy was in many cases the starting point of EU/EEC environmental policy: Germany introduced substantial environmental protection legislation earlier that other EU/EEC countries. This resulted in calls from German business for conditions of equal competition. They demanded that competitors in other Member States should be subject to the same environmental requirements and thus the same presumed costs. (Von Seht & Oht, 2000:5)

As discussed in previous chapters much of Germany's environmental policy-making has resulted from a discursively inclusive process. The above argument points to discourse continuing to be important in the transfer of environmental policy-making to the European level, i.e. Germany business communities entered into discourse regarding the need to prevent themselves being put at an economic disadvantage. Calls for EU/EEC legislation on the environment would undoubtedly have also come from actors promoting environmental awareness and protection for the environment's sake, thus resulting in multi-participative discourse.

Following the decision in 1972 to consider environmental concerns, the European Council adopted their First Action Programme on the Environment in 1973. Primary objectives that were identified in this first programme were for EEC environmental policy to

> i. prevent, reduce and as far as possible eliminate pollution and nuisances;
> ii. maintain a satisfactory ecological balance and ensure the protection of the biosphere;
> iii. ensure the sound management of and avoid any exploitation of resources or of nature which cause significant damage to the ecological balance;
> iv. guide development in accordance with quality requirements especially by improving working conditions and settings of life;
> v. ensure that more account is taken of environmental aspects in town planning and land use;
> vi. seek common solutions to environmental problems with States outside the Community, particularly in international organisations. (Commission of the European Communities, 1984:18)

These are grand-scale objectives that can still be seen to influence the way in which contemporary environmental policy is conducted. This will be evident when discussing recent and current policy.

Previously mentioned was the catalytic effect that *Waldsterben* had on German domestic environmental politics in the early 1980s. This transferred to Germany's efforts in Europe where they pushed for air pollution policies to be introduced. Anderson and Liefferink (1997:26) argue that Germany's efforts were aided by the appointment of a German Environment Commissioner, Karl-Heinz Narjes and Germany holding the Presidency of the Council in the first half of 1983, which had the effect of expediting a final proposal of an acidification policy that the commission was in the process of preparing. It is also pointed out that at this time the Third Environmental Action Programme was approved which placed more importance on the separation of environmental issues from trade concerns than previous programmes had done. A similar argument is made by Economy and Schreurs (1997) who cite the success of the Green Party in Germany, and the subsequent uptake of green issues by mainstream political parties (as discussed in chapter three) arguing that this train of events continued on to the European level where

> Germany became a "primary force" in pushing the European Community on the introduction of a Large Combustion Plant Directive and later in the cases of stratospheric ozone depletion and global climate change. (Economy and Schreurs, 1997:9)

Anderson and Liefferink state that 'Germany became known as the 'engine' of EU environmental policy in the 1980s' (1997:26). Reasons given for this are: Germany's domestic policies; the economic importance of the German market; and 'the more intentional pusher role of the German government' (Ibid).

Although EEC environmental policy-making became established in the 1970s and 1980s

> It was not until the Single European Act amended the Treaty of Rome in 1987 that express authority for an environmental policy was provided, thus effectively legitimizing the extensive body of environmental legislation that had by then been adopted under a rather elastic interpretation of the original Treaty. (Haigh, 1996:159)

Clearly the Member States of the EEC had agreed on the general development of environmental policies; the de facto existence of such policies eventually leading to their legitimisation and the right to introduce future policies.

The Single European Act decrees that a number of environmental principles be followed. These principles determine that preventive action should be taken to avoid environmental degradation; that environmental damage be addressed at source as a matter of urgency; that the polluter should pay; that environmental protection requirements be incorporated within other EEC policies; and that the principle of subsidiarity should apply. This latter principle means that action should be taken by the EEC only where the required objectives can be better achieved by the EEC than by Member States acting alone.

On the 1 November 1993 the Treaty on European Union (otherwise known as the Maastricht Treaty) came into effect. The treaty strengthened the requirement for environmental protection to be integrated both in definition and in the implementation of other policies, i.e. that the environment must be considered in all policy-making, not only that which is purely environmental. The environmental principles that were enshrined in the Single European Act were supplemented by the addition of the precautionary principle. Germany actively promoted its adoption. It is pertinent to mention that the precautionary principle is not one that sits easily with all countries of the EU. As Wurzel (2002) explains, a standard setting philosophy in Britain is the requirement for scientific proof. This fact indicates that for the precautionary principle to be adopted by the EU required discourse through which agreement was reached by the force of the better argument. The preventive principle and the precautionary principle may seem to be similar; the difference being that one can take preventive action if an outcome is expected, i.e. known to be an end product of a particular action or series of actions, the precautionary principle can be explained as requiring consideration of what outcomes MAY occur as a result of an action or series of actions and to take precautions accordingly.

The European Union consists of three pillars, which are the European Community (EC), Common Foreign and Security Policy, and Home Affairs and Justice Policy. The Maastricht Treaty amended the Treaty of Rome, which is the pre-existing pillar of the EU, changing its name from the EEC to the EC (europa.eu.int/abc/treaties_en.htm November 2004; Haigh, 1996). Whilst it is common practice to use the term EU when talking or writing about European law, it is in fact the case that it is only the EC that has legal identity. It is therefore, the case that the United Nations Framework Convention on Climate Change and the Kyoto Protocol are signed and ratified by the EC and the Members States. The 'legislature' of the EC consists of the Commission, the Council of Ministers and the Parliament. In addition to these bodies the Court of

Justice and the Court of Auditors complete the EC. The Commission is made up of twenty Commissioners and thirty-six Directorate Generals and specialised services. Margot Wallström was the Commissioner responsible for the Environment between September 1999 and November 2004, thus during preparation for, and at the time of, the WSSD and COP8.

Functions of the Directorate General for the Environment include: -

- Initiating new environmental legislation. Before issuing draft legislation, "preliminary soundings and discussions with representatives of governments, environmental NGOs, industry, special interest groups and, where necessary, technical experts." (DG Environment, 2002:4) are carried out. "Once a Commission proposal has been submitted to the Council of Ministers and the European Parliament, the three institutions work together to produce the final text." (Ibid:5)
- Ensuring that Member States apply environmental legislation correctly. Action can be taken against those in the both the public and private sectors that do not meet treaty obligations. With this in mind, it is important to note that EU directives can nevertheless be interpreted within the policy styles of individual Member States.
- Representing the EU at the international level and promoting international action to address global and transboundary environmental problems.Working towards the integration of environmental concerns in other policy areas.

DG Environment works within a framework set down by Environmental Action Programmes. The first of these programmes to specifically mention climate change other than requiring more research, was entitled 'Towards Sustainability' and was the Fifth Environmental Action Programme; it was approved on 1 February 1993. Amongst the programme's aims were: to ensure that all areas of policy-making consider environmental issues; to ensure the involvement of those areas causing environmental degradation; and to involve all relevant actors i.e. not just governments but to have an inclusive policy. This approach owed much to the lead taken by the Dutch National Environmental Policy. However, climate change issues were considered during the previous framework period. The European Parliament passed a resolution in 1986 that stated that climate change should be included in EC policy, but it was only in October 1990 that any political agreement was reached regarding any specific policy, this was that CO_2 emissions be stabilised at 1990 levels across the EC by 2000. It is also the case the the EC was present at the United Nations Conference on Environment and Development in 1992 and that the EC signed the United Nations Framework Convention on Climate Change.[1]

In June 2000 and thus during the period of the Fifth Environmental Action Programme, the European Climate Change Programme (ECCP) was launched, its aim being to create a strategy that enables the EU to implement its commitments under the Kyoto Protocol. The policy-making process within the ECCP has been multi-participatory with representatives from the various Directorate Generals, Member States, industry and environmental groups being included in consultative groups. It should also be mentioned that as well as Member States contributing to EU policy-making, this process is sometimes reversed. For example, the EU influences Germany in terms of policy learning, this can be seen in the defining of standards for clean air and regulating emissions.

The Sixth Environmental Action Programme runs from 2001 to 2010. Four priority areas to be concentrated upon are identified. These are: - climate change; nature and biodiversity; environment, health and quality of life; and natural resources and waste.

Five Major Objectives are identified, these are to: -

- Improve the implementation of existing environmental legislation at national and regional levels.
- Integrate environmental concerns into other policy areas.
- Work with business and consumers in a more market-driven approach to identify solutions.
- Ensure availability of better and more accessible information for citizens.
- Develop a more environmentally conscious attitude towards land-use planning.

This section has given a brief overview of the evolution of climate change related policies of the EU. The following section will examine how an EU position for climate change international relations is formed.

International Relations and the EU

The EU presents a united position at global conferences such as the WSSD and at climate change specific international negotiations such as those held annually by the Conference of the Parties to the UNFCCC. The EC (represented by the DG Environment) and the EU Member States attend such conferences, but it is the 'troika', which consists of the current EU Presidency, the incoming Presidency and the European Commission, that negotiates on behalf of all Member States. The presiding EU Presidency presents the pre-agreed united position. Speaking with 'one voice' gives the EU a strong bargaining position in terms of it representing a large power bloc. In recent years, in addition to the fifteen Member States, it has usually been the case that the ten European States that became EU members in May 2004 have also backed the EU position.

A bloc of twenty-five countries that are presenting a unified argument have a greater chance of influencing events than does any individual European State acting alone. It can be argued that without the proactivity of the EU following the withdrawal of the USA from the Kyoto Protocol, it is likely that the Protocol would have completely collapsed. Ratification of the Kyoto Protocol by Japan, the EC and the EU Member States meant that when Russia ratified the Protocol on 18 November 2004, the required ratification by 55 parties to the convention making up at least 55 per cent of 1990 emissions of Annex 1 parties (industrialised countries) occurred. The Protocol entered into force on 16 February 2005.

There are, however, some disadvantages of 'speaking with one voice'. Whilst the bargaining position of the EU as a bloc is enhanced in terms of pure power, flexibility suffers as does the ability to network, which would enable greater understanding of the positions of others and the chance to let others know and understand one's own position and perhaps gain influence through reasonable and just argument.

Some elucidation regarding the comment about flexibility is required. Prior to international negotiations, EU members agree on a unified stance. During international conferences EU coordination meetings are held at the start of each day so that the EU Member States can keep up to date with events and agree on a unified position. Coordination meetings can be, and often are, held during the course of the day to ensure that the EU 'one voice' keeps apace with developments. When discussions move on a tangent that was not expected, the EU tends to be left out of the ongoing debate whilst the EU Member States gather to talk about the new situation in order to come to some agreement amongst themselves before returning to the bargaining table. Thus, the EU can be rather slow at international negotiations; this situation is likely to be exacerbated now EU enlargement has occurred. Negotiators involved in these meetings are themselves aware of these shortcomings and are looking at ways to circumvent such problems. It could be argued that whilst discussions occurring between EU Member States in order to reach agreement may fit within Habermas' discourse ethics framework, that when this occurs repeatedly during international conferences this hinders the fulfilling of universally inclusive ethical discourse between the wider international community.

It is perhaps worth mentioning the argument put forward by Schelling (1960:29) that when a bargaining process has a time limit and there is seen to be inflexibility on the part of an actor that this can in fact work as an incentive to others to agree on a position near to that put forward by the inflexible actor.

The second disadvantage identified above, is the restricted time spent networking due to discussions being held among EU Member States. In an interview with Karsten Sach[2] of the BMU and a senior member of the

German delegation, what has been termed networking in this article was called outreach. Sach stated that this problem has been recognised and that at recent meetings it had been decided to maximise outreach by allocating certain areas to the most appropriate EU Member States; for example, France had the task of talking to representatives from francophone Africa, and Britain to Indian representatives. It will be seen in the following chapters that Germany is active in promoting outreach talks.

In order to arrive at a unified EU position prior to international negotiations, delegations from Member States meet regularly together with representatives of the European Commission (actually from the Environment Directorate General of the European Commission). Normal practice is for meetings of the 'working party on international environmental issues – climate change' to be held approximately monthly. These working party meetings are official meetings where preparations for Council meetings are conducted, together with the preparations and coordination of the positions to be taken at the international level. Additionally, expert groups are held to look into specific issue areas; these expert groups are initiated by, and report back to, the working party. Member States can appoint an expert to attend these groups and to accompany governmental representatives to conferences. In addition to the monthly meetings, at the start of each Presidency (January and July) a meeting is held that lasts for about three days. These meetings combine work and the reinforcing of working relationships; it is a time when informal ideas can be talked about. It is also the case that at least some of people working on climate change within the various countries and the European Commission have built up a close working relationship with one another; it can be that e-mails are exchanged frequently between various parties, for instance a close working relationship and regular contact exists between Germany and the UK.[3] In addition to the working party on climate change there is a 'working party on international issues – sustainable development'. It is also the case that just as Germany's climate change policies have evolved over a number of years so have those of the EU. The position taken by Germany in the climate change working groups was largely the same as that taken by Germany at the WSSD and COP8,[4] most of which was discussed in the previous chapter and which will be further examined when analysing these conferences. Continuing discourse exists between the Member States with the aim of reaching a unified position. Germany may have influenced and also been influenced by these discussions but the end result is that there is largely a confluence of positions. Some pertinent points with regard to the development of EU policy made during interviews include that made by: Sebastian Oberthür[5] who commented that Germany has heavily influenced European climate change policy; and Bill Hare[6] who stated that Germany is "definitely one of the, if not the leader in international climate negotiations and significant though not to an absolute degree with the EU". It thus appears that the flow of influence is mainly from Germany to the EU.

The process of arriving at an EU position is multi-layered, complicated and not necessarily linear; discussions and agreements need to occur at the domestic level, and then EU level, recommendations and considerations may then be referred back to the domestic level before being brought back to the EU discussions. Negotiations are not always purely between the EU and national levels, for example, jointly held positions of the DG Environment and the BMU may initially differ from the stance taken by the DG Industry and the BMWA. Theoretically, the negotiating process could continue for quite some time. Minutes are not taken at the EU negotiating meetings; it is therefore, difficult to ascertain the exact course of events. The outcomes of such meetings can, it appears, only be determined by looking at EU inputs to international negotiations.

Not all EU Member States attend all meetings, not all States have the resources either human and/or financial to be able to do so. Germany does have the resources and does choose to participate at meetings. Those States that do not attend meetings are clearly disadvantaged vis-à-vis the larger EU States, as the former miss out on the opportunity to influence the EU position. Nevertheless, it can be argued that these resource poor States are advantaged by the very fact that they are included in the EU; they have after all the EU 'power bloc' to argue their case (albeit they may not have had much input into the EU standpoint). If they were outside of the EU they would still be resource poor and perhaps not be able to participate fully in international negotiations. They are therefore, possibly advantaged vis-à-vis non-EU small States. It is also the case that these poorer States of the EU can prioritise the issues that they wish most to participate in and thus save their resources for issues of their choice. They can also keep track of the EU stance on matters they are not directly participating in, the EU's position could conceivably coincide with their own anyway, they can conserve their own resources in the knowledge that the EU will in many ways be looking after their interests. It is also the case that some of the poorer countries within the EU have the kudos of having ratified the Kyoto Protocol but are able to increase their own GHG emissions as the EU burden sharing agreement means that the EU must reduce its emissions by 8 per cent but this can be internally distributed. Germany for example has to reduce its emissions by 21 per cent whilst Portugal can increase by 27 per cent.[7]

The procedure through which an EU position for international climate change related negotiations is developed has been examined in this section. Influence that Germany has had in relation to the EU's objectives for the WSSD will now be considered.

Germany's contribution to the EU's aims for the WSSD

As already stated minutes are not taken at the meetings between Member States and the European Commission, through which the 'EU one voice' is formulated. Available documentation and academic literature concentrates on either the State level or the EU level but not on the interaction between the two. The lack of academic literature regarding this interaction can be explained by the absence of meetings minutes. Influence that any one Member State may have within the EU needs to be assessed through comparison of State aims and those of the EU, together with interviews with a variety of parties.

When discussing international relations, a point that persistently arose during interviews conducted in Germany was that it is 'Germany's way' to try reach consensus through discourse and that this is largely due to Germany's history. The historic need for Germany to be 'contained' also arose, resulting in Germany now traditionally working through and within Europe.

Germany has representatives at all meetings, it has the resources both financial and human to be able to do so; this is not the case for all EU countries. States that are able to participate fully in all meetings clearly have an advantage in getting their views incorporated into the EU position; in other words, they are more likely to be able to influence the position that is adopted by the EU. Not only is Germany represented at all meetings, but it also takes an active part in them. It has been said[8] that Karsten Sach of the BMU is a very active, well-respected and authoritative member of the working group, and that people listen to him.

The lack of transparency in the formation of the EU's position makes it extremely difficult to prove a causal link between Germany's position and that of the EU. Nevertheless, a number of interviewees (these include people from Germany, the EU Directorate General for the Environment and the UK's Department of Environment, Food and Rural Affairs - DEFRA) confirmed that Germany has been active and influential within the EU with regards to environmental matters and specifically with regards to climate change policy, and that Karsten Sach provides an important part of the German influence. Other reasons given in the DEFRA interview[9] for Germany's influence within the EU were the contribution in monetary terms that Germany makes to the EU, and its delivery on targets, i.e. its successful domestic policies. Germany's proactive climate change related domestic policy-making gives credence to Germany's arguments, both within the EU and the international arena. The fact that the EU's ability to reach its Kyoto Protocol target of reducing greenhouse gas emissions is largely reliant on Germany's performance is evidence of the importance of Germany within EU policy making regarding climate change.

It should be noted that States other than Germany have also exerted influence at various junctures, most notably the Netherlands, Denmark and the UK.

With regard to specific policies an article placed on the German government's website on 23rd August 2002 states that

> The EU wants to present four strategy objectives in Johannesburg:
>
> • More fairness worldwide and more effective partnership for sustainable development.
>
> • Stronger integration of environment and development at the international level.
>
> • Agreement on environment and development objectives to revitalize and strengthen the Rio process.
>
> • More effective international monitoring of national strategies. (www.bundesregierung.de/en/News-by-subject/International-,11075/Summits-and-Conferences.htm August 2002)

These four strategies clearly coincide with Germany's aims. It is intimated that discourse should be entered into that includes justice considerations. This is reflective of the strategy that Germany pursues. It can be argued that this aim is a step in the right direction, and one that is necessary if implementation is to occur. However, implementation of the ideals themselves is another matter and will be the subject of discussion in the following chapters.

Other main points of the EU stance include:

> All countries must work together, recognising their common but differentiated responsibilities, to ensure that growth is decoupled from environmental degradation and that the needs of the present generation are satisfied without destroying the capacity of future generations to cater for their needs.
>
> …
>
> To enhance the use of cleaner, more efficient fossil fuel technologies, to improve energy efficiency and to increase the share of renewable energy sources to at least 15% of primary energy supply by 2010.
>
> …

> To develop a ten-year work programme to accelerate the shift towards sustainable consumption and production. (European Commission, 2002a)
>
> The EU will push hard for an action plan that specifies targets, timetable and schemes of work. (European Commission, 2002b: 11)

Again it can be seen that these points are reflective of Germany's aims.

As previously mentioned, agenda setting is one of the facets of power as outlined by Lukes (1974). An example of German and EU influence in this respect relates to energy and the WSSD agenda. According to Andreas Krämer[10] of Ecologic, the German government wanted energy to be on the agenda at the WSSD but that energy is not a part of Agenda 21, and that in the WSSD PrepComs it was not clear that energy would be on the agenda. The EU, led by France and Austria pushed for energy to be included; this initiative was strongly supported by Germany. Ecologic looked into some ideas regarding this issue, which were then taken up by the BMZ, who hosted a conference that was attended by, amongst others, Klaus Töpfer of UNEP. Shortly thereafter energy was included on the agenda for the WSSD. The EU's initiative was strengthened by the BMZ's conference and the presence of Töpfer; Germany can be seen to have contributed to the WSSD agenda setting. This example demonstrates the transferability of ideas, through discourse between various political levels.

As previously discussed, Germany would like to see UNEP transformed into a World Environment Organisation, which would be an effective counterweight to the WTO. The EU's position did not include advocating creation of such an organisation. The EU did, however, promote improved governance at all levels in order to successfully implement sustainable development programmes.

> At international level it is necessary to strengthen United Nations' bodies such as the Economic and Social Council and the Commission on Sustainable Development to ensure that the results of the World Summit are followed-up and implemented properly. (European Commission, 2002b:11)
>
> and to reinforce co-operation on sustainable development between UN bodies, the Bretton Woods institutions and the WTO. (European Commission, 2002a)

Germany's position and that of the EU were not identical. Nevertheless calls for the strengthening of international bodies were made, as was increased cooperation between UN bodies and the WTO. If Germany had any influence at all in the formation of this particular strategy, it was not complete. It is perhaps worth mentioning that within the EU's sixth Environment Action Programme the strengthening of UNEP is advocated.

With reference to COP8, both Peter Fischer[11] from the AA and Karsten Sach[12] from the BMU confirmed that Germany wanted dialogue to commence as soon as possible with regard to the future commitment period of the Kyoto protocol (from 2012). This was also an objective pursued by the EU. It is safe to assume that Germany would have been at the very least a supporter of this policy within the EU, and may have had an influential role in the development of the EU position.

More of the EU position will become apparent in the next chapter, in which events at the WSSD and COP8 are analysed. It will be seen that the issues advocated by the EU are largely synonymous with those already identified by Germany as aims. It is impossible to say to what degree Germany has definitely influenced the EU position; it is probable that in many instances influences are reflexive, i.e. they travel in both directions. However, it has also been seen that many of the aims taken on board by the German government have been called for by a variety of bodies within Germany. It is a feature of decisions that have been reached through multi-participative discourse that it is difficult to determine where influence lies. It is probable that multifarious factors and influences exist.

Conclusion

It has been argued that Germany actively participates in the evolution of environmental policy within the EU. The very nature of the EU, with Member States regularly meeting to discuss environmental issues in order to reach a united position can be related to Habermas' requirements for ethical discourse. That said some of the reasons cited for Germany's influence, i.e. monetary contributions and market importance do not fit within the framework of the force of the better argument prevailing. However, the fact that existing policies within Germany give credibility to the German position can be related to this framework, as if a policy is already seen to be working efficiently then one could say that this is indicative of the best argument. The fact that Germany actively participates in all meetings and is keen to enter into discussions with all parties also fits within the requirements for ethical discourse.

There is a large degree of correlation between Germany's aims and those of the EU, and it is widely believed that Germany does exert influence in the EU's international policy formation process. This influence is achieved through a variety of means such as: the personal agency of Karsten Sach, and structural issues such as the fact that Germany is a large monetary contributor to the EU and credibility of Germany's argument due to its successful domestic policies. Chapter four included the assertion by both Fues and Trittin that Germany's domestic policies give credibility to the stance it takes in international relations. This chapter argues that this is a contributing factor to Germany's influence within the EU.

It has been argued that Germany is influential with regard to EU climate change policies. It should be noted that it is also the case that if a policy is backed by the EU, it reinforces the BMU's argument within Germany. There is reflexivity between German and EU policies and aims.

Not all of Germany's aims have been taken up by the EU. One of the features of decisions reached through multi-participative discursive processes can be that where true consensus is not achieved compromise is needed. In order to arrive at the united position that the EU presents at international negotiations, the EU Member States have to negotiate and to varying degrees they have to compromise. Germany is one of the more influential member States; nevertheless, the EU's position is not Germany's per se, for example Germany's aim of UNEP being transformed into a powerful WEO was not a part of the EU's position at the WSSD. Thus Germany's aims, at least in so far as it acts through the EU are mediated by international factors.

Previous chapters have argued that analysis of Germany's climate change politics using Habermas' work on discourse ethics is appropriate due to the discursively inclusive nature of environmental policy development in Germany. It can be argued that as the structure of the EU necessitates participative justice as far as its Member States are concerned, and that as German participants are well versed in such processes, they could be seen to be at an advantage due to these competences. It can also be argued that EU delegations at truly international meetings have similar advantages vis-à-vis States that are not regularly a part of discursively inclusive processes or at least not to the same degree as do climate change negotiators within the EU.

This chapter has examined the way in which environmental policies in the EU have evolved and the way in which Germany engages with EU policy-making. The following chapter will assess international climate change related negotiations at the WSSD and COP8.

6

WSSD AND COP8

Previous chapters have examined the evolution of climate change related environmental policies and politics, specifically within Germany and more generically at the international level. It has been shown that inherent in German politics and in the formation of Germany's climate change related policies is the propensity for discourse that is inclusive of a cross-section of the population. It has also been shown that Germany has successfully implemented a number of domestic policies, such as the promotion of renewable energies. Such policies are beneficial domestically and they also give credibility to Germany's arguments when negotiating in the international arena. Germany and the EU aimed for renewable energy targets to be agreed on at the WSSD. This chapter will show that although this aim was not fulfilled, Germany and the EU still managed to influence international negotiations by initiating a coalition of like-minded countries that would continue negotiations aimed at committing to targets.

This chapter looks at events at the WSSD and COP8 and assesses Germany's influence at these events. It should be remembered, that when mention is made of an EC/EU position, it is a unified stance of EU Member States and is therefore, also representative of Germany.

As this chapter is read the following quote from Habermas is worth keeping in mind

> Under the pragmatic presuppositions of an inclusive and noncoercive rational discourse among free and equal participants, everyone is required to take the perspective of everyone else, and thus project herself into the understandings of self and world of all others; from this interlocking of perspectives there emerges an ideally extended we-perspective from which all can test in common whether they wish to make a controversial norm the basis of their shared practice; and this should include mutual criticism of the appropriateness of the languages in terms of which situations and needs are interpreted. In the course of successfully taken

abstractions, the core of generalizable interests can emerge step by step. (Cavalier *et al,* 2002, quoting Habermas, 1995:117-118)

It is with these aims in mind that it is hoped participants of international environmental negotiations enter discussion.

Prior to examining the proceedings of the conferences, some background information on the preparatory process of the WSSD will be given. The WSSD was held on 26 August–4 September 2002, prior to this and integral to the WSSD process, Sessions of the Preparatory Committee for the WSSD (PrepComs) were held. There were four PrepComs, all of which were a part of the tenth session of the Commission for Sustainable Development (CSD). The first three PrepComs were held at the UN in New York: the first was held on 20 April–2 May 2001; the second on 28 January–9 February 2002; and the third on the 25 March–5 April 2002. PrepCom IV was the culmination of the official pre WSSD process and at which, it was hoped much of the agreement to be reached at the WSSD would be decided upon. PrepCom IV was held in Bali on 27 May–7 June 2002, prior to which there were two days of informal consultations. In addition to the PrepComs, there were regional Eminent Persons' Roundtables on the WSSD, plus regional preparatory meetings for Europe and North America, which took place on 24–25 September 2001; for Latin America and the Caribbean; West Asia; Asia and the Pacific; and the last one for small island developing States held on 7–11 January 2002. The G8 Environmental Ministers Meeting in Banff, Canada on 12–14 April 2002 also addressed issues to be dealt with at the WSSD. In the run-up to the WSSD, many discussions would have been held at a variety of meetings and most probably between everyday network contacts. It is not possible here to analyse all such discussions, but it is worth noting that these occur and that events at the WSSD and COP8 are the culmination of a multitude of inputs. As pointed out previously in this study, global summits can produce an impetus for the advancement of related policies. It is possible that the then upcoming WSSD may have influenced the decision of the GEF Council to approve US$113.7 million for twenty environmental projects at the Third Replenishment of the GEF Trust Fund meeting in Washington D.C. held on 13–14 May 2002.

WSSD

The first and major part of this section examines the formal documentation resulting from the WSSD. Following on from this will be analysis of: events and discussions that were extra to the formal process; the speech given to the summit by Gerhard Schröder; and other efforts by the German delegation. The second part of the section will compare the results of the WSSD with German and EU aims.

Climate change related issues at WSSD

World Summit on Sustainable Development: Plan of Implementation

I. Introduction

The WSSD Plan of Implementation begins by reaffirming commitment to the principles agreed at the 1992 United Nations Conference on Environment and Development (UNCED) (thereafter these are referred to as the Rio principles), to the full implementation of Agenda 21 and to internationally agreed development goals.

The intention of the Plan of Implementation is to

> Further build on the achievements made since UNCED and expedite the realization of the remaining goals. To this end, we commit ourselves to undertaking concrete actions and measures at all levels and to enhancing international cooperation, taking into account the Rio Principles, including, inter alia, the principle of common but differentiated responsibilities set out in Principle 7 of the Rio Declaration on Environment and Development. (WSSD Plan of Implementation, 2002)

The principle of common but differentiated responsibilities is seen by many as being fundamental, it is of great relevance to how problems are perceived and therefore, to how solutions are worked towards. This principle is extremely pertinent in the international political climate change arena (the UNFCCC was born out of the UNCED process). It is worth pointing out that during the WSSD preparatory process the USA, Japan and Australia wanted reference to this principle to be deleted (International Institute for Sustainable Development, 2002, Vol.22. No.30:6). The three countries that opposed this principle are large, the USA and to a lesser extent Japan are normally associated with having considerable power. Clearly for this principle to have been kept in the face of opposition from such powerful quarters, the vast majority of countries must have opposed its deletion. This is an example where the discursive principles and inclusiveness that are applied in international negotiations and that are akin to those espoused by Habermas have led to a decision that is perceived as being just. Germany is amongst those countries that support the principle of common but differentiated responsibility.[1]

The Plan recognises the need to respect human rights and cultural diversity, it is stated that these are amongst the requirements for sustainable development to be available to all people. The link to climate change may not be obvious; however, it can be argued that it is a human right to be able to live in one's homeland with access to clean drinkable water and to food. An example of where climate change is threatening these rights can be found in some of the Pacific Islands, which are experiencing salt-water inundation of groundwater supplies; this threatens access to drinking water and also prevents the growing of some food stuffs. It is also the case that some islands are losing land to encroaching seas. Tuvalu is already trying to gain agreement for its population to migrate due to problems resulting from climate change. The need to respect human rights and cultural diversity was proposed for inclusion in the Plan of Implementation, during the preparatory process, by the EU (International Institute for Sustainable Development, 2002, Vol.22.No.30:6). The proposal would have been discussed and consensually agreed upon, thus leading to the link between human rights, cultural diversity and universal sustainable development being enshrined in an internationally agreed document. It can be seen that through multi-participative discussions norm construction can be achieved. It can be argued that the implementation of Habermasian discourse ethics, where all effected people have the right to participate and thus to have their concerns considered, would be a step toward respecting human rights, and that the process just described could be a first step towards that objective.

II. Poverty Eradication

Access to reliable and affordable energy sources is sought in the aim to eradicate poverty. The need for and the intention to improve energy supplies that are affordable, socially acceptable (presumably not nuclear) and environmentally sound are discussed. Increased use of renewable energy is called for, as is increased efficiency in energy consumption. In order to reach these aims capacity building and financial and technical assistance are called for, as are regional and international cooperation in support of national efforts. It is envisaged that public-private partnerships be included in the provision of financial and technical assistance given by developed countries to those in need. These ideals are championed within and by Germany. Previously in this study it was pointed out that Germany and the EU were influential in getting energy onto the WSSD agenda, i.e. enabling its inclusion in multi-participative discourse. Germany and the EU also wanted renewable energy targets to be introduced. Germany actively promotes renewable energy and energy efficiency. Whilst definitive amounts of influence cannot be apportioned to Germany with regard to recognition of the need for increased use of renewable energy and energy efficiency, it can be seen that Germany did play a part in getting the issue discussed. It can be argued that Germany used its expertise in multi-participatory discursive processes to enable this subject to be discussed and that its inclusion in the Plan was the result of Habermasian communicative rationality.

III. Changing unsustainable patterns of consumption and production

In order to achieve sustainable development the ways in which production is achieved need to be assessed and in many cases changed, as do patterns of consumption. Developed countries are those that produce and consume most, therefore it is these countries that need to take the lead in introducing these changes. Again this section refers to the Rio principle of common but differentiated responsibilities. Point fourteen reads:

> Encourage and promote the development of a 10 year framework of programmes in support of regional and national initiatives to accelerate the shift towards sustainable consumption and production to promote social and economic development within the carrying capacity of ecosystems by addressing and, where appropriate delinking economic growth and environmental degradation through improving efficiency and sustainability in the use of resources and production processes, and reducing resource degradation, pollution and waste. All countries should take action, with developed countries taking the lead, taking into account the development needs and capabilities of developing countries through mobilization, from all sources, of financial and technical assistance and capacity-building for developing countries. (WSSD Plan of Implementation, 2002)

The document goes on to detail various required courses of action for example; awareness raising programmes and the need to apply the polluter-pays principle (principle 16 in the Rio Declaration). At PrepCom IV in Bali the terminology to be used was contested; economic development was favoured by the EU, Hungary and the G77/China, whilst the USA and Australia wanted to use economic growth. It is stated in an Earth Negotiations Bulletin report that delegates accepted 'economic development' (International Institute for Sustainable Development, 2002, Vol.22.No.30:6). As can be seen in the above paragraph social and economic development is to be sought, whilst economic growth must be de-linked from environmental degradation. It may appear a matter of semantics but the use of these terms in the particular situation in which they have been used seems to be a subtle form of changing the way in which these terms are perceived. Whilst the message is that economic growth is acceptable if it does not incur environment degradation, the linkage between the all out pursuit of economic growth and the environment being damaged is made. The correct use of language and its contribution to people's understanding of discourse is recognised by Habermas (1979), who argues that both linguistic competence and communicative competence are necessary to achieve universal understanding. The use of language is important in norm construction as can be seen by the attempt to endow the term 'development' with positive links and 'growth' with the notion that unregulated growth can be damaging. The EU was active in the promotion of the term 'social and economic development'. It is also worth noting that during the Bali PrepCom it was decided that the EU lead an informal group to consider the development of a ten-year programme to improve resource efficiency.

Point fifteen of the Plan details the need for increased

> investment in cleaner production and eco-efficiency in all countries, through, inter alia, incentives and support schemes and policies directed at establishing appropriate regulatory, financial and legal frameworks. (WSSD Plan of Implementation, 2002)

The documentation does not state the origins of this point nor the supporters of it. However, domestic policy within Germany can be seen as leading the way, the electricity feed-in laws being a prime example. As explained in chapter three, these laws require utility firms to connect renewable energy supplies and combined heat and power supplies to the grid and to purchase the power at a premium price.

Point nineteen calls upon governments and regional and international organisations to implement the recommendations and conclusion of the ninth session of the Commission on Sustainable Development regarding energy and sustainable development.[2] Some of the points detailed in the Plan are the need for: diffusion of environmentally sound technologies; considerations regarding energy efficiency, affordability etc. to be included in the running of all manner of industries; development and dissemination of new energy technologies which increase the usage of renewable energy and improve efficiency of usage. The call for diversification of energy supply sources, development of cleaner and more energy efficient technologies and the need of such developments to be made available to developing countries can be related to of the role of the *Deutsche Gesellschaft für Technische Zusammenarbeit* (GTZ). This is a German government owned corporation that invests in sustainable development programmes in developing countries. Clearly it is not only Germany that runs such programmes, nevertheless, it is worth noting that Germany does already aspire to and work towards the aims set out in the Plan of Implementation. Again, a correlation of aims can be seen, even though influence cannot be definitively attributed to Germany.

Also detailed is the need to ensure that market policies encourage energy systems that are compatible with sustainable development; to this end improved market signals, the removal of market distortions, restructuring of taxation and the phasing out of harmful subsidies are advocated. The call for improved market signals and the removal of market distortions is somewhat ambiguous and therefore, open to interpretation. However, the feed-in laws in Germany that encourage renewable energy production are examples of improved market signals. It could be that Germany's effective domestic policies with regard to the development of renewable energy have been taken note of, and have given credibility to, the call for such policies at a global level. In Habermasian terms this can be translated as a proven 'best' argument prevailing in international policy-making discourse.

Also called for is strengthening of

> national and regional energy institutions or arrangements for enhancing regional and international cooperation on energy for sustainable development, in particular to assist developing countries in their domestic efforts to provide reliable, affordable, economically viable, socially acceptable and environmentally sound energy services to all sections of their populations; (WSSD Plan of Implementation, 2002).

Also advocated is the promotion of international and regional cooperation towards such aims, public/private partnerships are specifically mentioned in this regard. Again Germany's GTZ can be held up as an example of something akin to that called for, already in practice. Other possible examples of such cooperation may prove to be the Coalition on Renewable Energy and the German hosted renewable energy conference that are discussed later in this section. As Germany was already active in promoting policies of the type advocated, it is likely that Germany would, at the very least, have supported the inclusion of this text in the Plan.

Point twenty addresses transportation; the same types of requirements needed to achieve sustainable development are identified as has been detailed above i.e. efficiency, environmentally sound etc. Transport specific requirements include the need to plan for public transport and goods delivery systems that reduce pollution, congestion, adverse health effects and greenhouse gas emissions. As has been detailed in chapter three, Germany is implementing policies to limit the use of roads for public goods deliveries and to encourage the use of public transport. Germany has also called for taxes to be imposed on aviation fuel. It is therefore, probable that when this point was being discussed, Germany would have supported its inclusion in the Plan.

IV. Protecting and managing the natural resource base of economic and social development

Point thirty six specifically addresses climate change, it reaffirms

> its adverse effects are a common concern of humankind. We remain deeply concerned that all countries, particularly developing countries including the least developed countries and small island

> developing States, face increased risks of negative impacts of climate change and recognize that, in this context, the problems of poverty, land degradation, access to water and food and human health remain at the center of global attention. The United Nations Framework Convention on Climate Change is the key instrument for addressing climate change, a global concern, and we reaffirm our commitment to achieving its ultimate objective of stabilization of greenhouse gas concentrations in the atmosphere at a level that would prevent dangerous anthropogenic interference with the climate system, ... States that have ratified the Kyoto Protocol strongly urge States that have not already done so to ratify the Kyoto Protocol in a timely manner. (WSSD Plan of Implementation, 2002).

Germany has ratified the Kyoto Protocol and would have been one of those countries pushing for the inclusion of this sentence in the Plan of Implementation. It is pertinent to note that during the Bali PrepCom Germany argued 'that the Kyoto Protocol is a manifestation of the precautionary principle, and that the deficit in implementation must be overcome' (International Institute for Sustainable Development, 2002, Vol.22.No39:4). Germany has been proactive in trying to get the Kyoto Protocol implemented; this point will be expanded on in the next chapter. It is also worth reiterating that the precautionary principle originated in Germany.

Required actions that are identified include:

> Actions at all levels are required to:
> (a) Meet all the commitments and obligations under the UNFCCC;
> (b) Work cooperatively towards achieving the objectives of the UNFCCC;
> (c) Provide technical and financial assistance and capacity building to developing countries and countries with economies in transition in accordance with commitments under the UNFCCC, including the Marrakesh accords;
> (d) Build and enhance scientific and technological capabilities, inter alia through continuing support to the IPCC for the exchange of scientific data and information especially in developing countries;
> (e) Develop and transfer technological solutions;
> (f) Develop and disseminate innovative technologies in respect of key sectors of development, particularly energy, and of investment in this regard, including through private sector

> involvement, market-oriented approaches, as well as supportive public policies and international cooperation; (WSSD Plan of Implementation, 2002).

The points following the above relate to actions required to continue observing climate change and its impacts. Point (f) above is again one that can be related to the work of the GTZ, in other words, this aim agreed by the WSSD is already in the process of being implemented by Germany. The fact that Germany is already engaged in the advocated policies means that it is highly probable that they would have been in favour of this aim being included in the Plan of Implementation. Yet again, influence by Germany can not be proven, although it is a distinct possibility.

V. Sustainable development in a globalizing world

Point forty five extols advantages of globalisation but it also recognises that 'there remain serious challenges' (WSSD Plan of Implementation, 2002) and that 'Globalization should be fully inclusive and equitable' (Ibid). Environmental damage is not mentioned as a consequence or 'challenge' of globalisation although this is often a problem; climate change is really the ultimate form of globalised environmental degradation. It will be seen later in this chapter that issues relating to equity in terms of the perception by many developing countries that industrialised countries are not taking responsibility for historical greenhouse gas emissions proved to be a problem in the COP8 negotiations. It can be argued that truly inclusive discourse with decisions made through communicative rationality would be a first step toward agreements that would address the problems of sustainable development in a globalising world.

VI. Health and sustainable development

This section acknowledges that 'There is an urgent need to address the causes of ill health, including environmental causes' (WSSD Plan of Implementation, 2002). Climate change can have a detrimental effect on people's health, especially in areas of the world where water supply is degraded, ground water is salinated thereby damaging food production, and also in the spread of diseases such as malaria to areas where previously climatic conditions were not conducive to mosquitoes. It can be argued that multi-participative ethical discourse is more likely to result in agreements that prevent practices that are damaging to human health than discussions that do not include those people whose health is endangered.

VII. Sustainable development of small island developing States

Small island developing States are recognised as leaders with regard to sustainable development, whilst being 'increasingly constrained by the interplay of adverse factors clearly underlined in Agenda 21' (WSSD Plan of Implementation, 2002). The adverse factors identified include being 'ecologically fragile and vulnerable' (UNCED, 1992:66) and economically disadvantaged due to their geographical remoteness. The Plan of Implementation identifies various actions that are required, these include:

> (j) Assist small island developing States in mobilizing adequate resources and partnerships for their adaptation needs relating to the adverse effects of climate change, sea level rise and climate variability, consistent with commitments under the United Nations Framework Convention on Climate Change, where applicable; (WSSD Plan of Implementation, 2002)

The Plan also expounds the need for environmentally sound energy supplies and states that existing efforts to supply energy should be strengthened and new efforts should be made by 2004. The development and promotion of energy efficiency, renewable energy sources and capacity building with regard to 'training, technical know-how and strengthening national institutions in the area of energy management' (WSSD Plan of Implementation, 2002) is also called for. It is perhaps, worth reiterating at this point, that the effects that Germany's influence in the international relations of climate change may or may not have on the low-lying Pacific Island countries will be examined in the next chapter.

VIII. Sustainable development for Africa

Required actions identified include the need to enable

> African countries to undertake environmental legislative policy and institutional reform for sustainable development and to undertake environmental impact assessments and, as appropriate, to negotiate and implement multilateral environmental agreements; (WSSD Plan of Implementation, 2002).

This section also addresses the need to provide support with regard to energy supplies; for efficient and cleaner usage of fossil fuels and increased use of renewable energies.

Action is called for to

> Assist African countries in mobilizing adequate resources for their adaptation needs relating to the adverse effects of climate change, extreme weather events, sea level rise and climate variability, and assist in developing national climate change strategies and mitigation programmes, and continue to take actions to mitigate the adverse effects on climate change in Africa, consistent with the United Nations Framework Convention on Climate Change; (WSSD Plan of Implementation, 2002).

Germany's GTZ concentrates its work in Africa, this, together with the fact that Germany makes available an expert to assist developing countries in carrying out environmental assessments through the UNFCCC, demonstrates that Germany is already implementing actions similar to those called for by the Plan of Implementation. This is an example where Germany is showing leadership. It is probable that given the combination of leading through action already being taken, and therefore, the almost certain support for the aims as detailed above, that Germany exerted some influence on the outcomes of these international negotiations.

IX. Means of Implementation

It is recognised that for the Plan, Agenda 21 and the various other international agreements that are reaffirmed in the Plan, to be implemented, requires increased effort by individual countries and the international community and increased availability of funds. The need for increased foreign direct investment and targeted overseas development assistance is identified. The funding of international organisations and agencies should be made more predictable and assured. This statement is reminiscent of calls by Trittin, for committed funding of a strengthened or transformed UNEP, that are detailed in chapter four. Whilst this does not denote influence on the part of Trittin, it does indicate that Germany is likely to have been advocating this policy. The third replenishment of the GEF is welcomed. GEF is encouraged to

> leverage additional funds from key public and private organizations, improve the management of funds through more speedy and streamlined procedures and simplify its project cycle. (WSSD Plan of Implementation, 2002).

Point one hundred and three calls for improved decision-making, with improved collaboration between scientists and policy-makers and consideration of scientific assessments. Partnerships between scientific, public and private institutions are also advocated. Action is called for to

> Promote and improve science-based decision-making and reaffirm the precautionary approach as set out in principle 15 of the Rio Declaration on Environment and Development, which states: "In order to protect the environment, the precautionary approach shall be widely applied by States according to their capabilities. Where there are threats of serious or irreversible damage, lack of full scientific certainty shall not be used as a reason for postponing cost-effective measures to prevent environmental degradation." (WSSD Plan of Implementation, 2002)

Multi-participatory collaboration as called for is to a large extent already in existence in Germany. The precautionary principle has been a foundation of German environmental politics since at least 1971. These are further examples of German leadership and possible influence. The precautionary principle was of course, agreed at UNCED and the high level segment of the PrepCom in Bali affirmed adherence to the Rio Principles. The last sentence of the above quote refers to 'cost-effective measures to prevent environmental degradation'; however, no definition is given for what is determined as cost-effectiveness and by whom? It can be argued that against the extremes of climate change most measures must surely be cost-effective.

Point one hundred and nineteen ter reads:

> Ensure access, at the national level, to environmental information and judicial and administrative proceedings in environmental matters, as well as public participation in decision-making, so as to further principle 10 of the Rio Declaration on Environment and Development, taking into full account principles 5, 7 and 11 of the Declaration. (WSSD Plan of Implementation, 2002)

Principle 10 states that 'Environmental issues are best handled with participation of all concerned citizens, at the relevant level. … States shall facilitate and encourage public awareness and participation…' (UNCED, 1992:2) Principle 5 refers to the need to eradicate poverty as an essential part of sustainable development; principle 7 refers to common but differentiated responsibilities; and principle 11 refers to enacting effective environmental legislation. All of these are at least being worked towards within Germany. The call for public participation in decision-making can be directly related to Habermas' call for everyone to be able to participate in decision-making discourse.

X. Institutional framework for sustainable development

Point one hundred and twenty two includes the need to

> (b) Strengthen the collaboration within and between the United Nations system, international financial institutions, the Global Environment Facility and WTO, utilizing the United Nations Chief Executives Board for Coordination (CEB), the United Nations Development Group, the Environment Management Group and other inter-agency coordinating bodies. Strengthened inter-agency collaboration should be pursued in all relevant contexts, … (WSSD Plan of Implementation, 2002)

One of Germany's long-standing hopes has been for UNEP to be strengthened to form a World Environment Organisation and for it to have the authority to balance the WTO. Clearly the above bears no resemblance to such a wish, however it is perhaps possible that the identified need for further collaboration between various environmental, financial and trade organisations is evidence of some minute influence from those calling for a strong environmental organisation. It could be that it would be unacceptable to many, including the WTO to actually use such terminology as the need to counter-balance the WTO and so extremely diplomatic language was used as a first step towards such an eventuality. It could be evidence of the power of those parties that place prime importance on trade and all matters financial, virtually no matter what the consequences are. It is of course, the result of the battle between these positions, or to explain it more diplomatically, it is the consensus or compromise achieved through multi-participatory discussions.

Point one hundred and twenty four states that

> A vibrant and effective United Nations system is fundamental to the promotion of international cooperation for sustainable development and to a global economic system that works for all. To this effect, a firm commitment to the ideals of the United Nations and to the principles of international law and those enshrined in the Charter of the United Nations, and to strengthening the United Nations system and other multilateral institutions and promoting the improvement of their operations, is essential. (WSSD Plan of Implementation, 2002)

Discussed above was the German desire for UNEP to be strengthened to become a WEO. Although the Plan of Implementation calls for the strengthening of the UN system, there is neither specific mention of UNEP nor any mention of some parts of the system being strengthened vis-à-vis others, i.e. to create equity of power. It should be noted that whilst not a part of the UN, the WTO is a related organisation according to the UN organisational chart. Clearly all participating countries agreed the Plan of Implementation, the USA included. One wonders how genuine the USA's commitment is, as its attitude toward UN bodies has in the recent past been variable, at times appearing to treat the UN with contempt.

The penultimate point and the last of the many quotations used, is it can be argued, of great moral significance. Point one hundred and fifty two reads

> Acknowledge the consideration being given to the possible relationship between environment and human rights, including the right to development, with full and transparent participation of Member States of the United Nations and observer States. (WSSD Plan of Implementation, 2002)

It can be argued that in some instances there most definitely is a relationship between the environment and human rights. In 2002 widespread flooding in central Europe, Germany included, led to the loss of life and homes. Some of the Pacific Islands are experiencing loss of land, of food producing capabilities and access to freshwater, mass emigrations due to necessity are foreseen. The most human basic rights, i.e. to life, to sustenance, and to a safe place to live are being denied. If discourse ethics as espoused by Habermas were followed on a globally inclusive basis, with the non-economic costs to people of GHG emissions being taken into account, as well as the possibilities to expand use of renewable energies being considered, one would hope that the human rights/environment linkage would not only be understood, but also that policies would be implemented to ensure that human rights were met rather than undermined. There is the argument that the world's poor need access to energy and conventionally produced energy is far cheaper and the technology more widespread than renewably sourced energy and therefore, the former should be supplied. In riposte, and in economic terms, if the true cost of conventionally produced energy (costs of freshwater and food supplies to remote areas of the world, of rebuilding costs, of migration costs, etc. - one cannot put a price on the loss of life) were internalised it would become vastly more expensive. If subsidies now given to conventional energy production were stopped and rerouted to renewable technologies these would become more developed, efficient and therefore, viable with regards to mass supply. Relating these comments to Germany, it is the case that coal subsidies are being decreased and renewable energy subsidies are being increased, however, the former still exceed by far the latter.[3]

Repeatedly mentioned throughout the Plan of Implementation are: the need to implement Agenda 21 and the need to adhere to the Rio principles such as common but differentiated responsibilities, polluter-pays and the precautionary principle. To reiterate, these latter two principles have been formally included in German decision-making procedures since 1971, and it can be argued that the common but differentiated responsibility principle is to an extent followed in practice.

The Johannesburg Declaration on Sustainable Development

The Plan of Implementation is the document of substance resulting from the WSSD. The Declaration basically serves as a briefing document on the principles agreed to, and challenges to be addressed. A few points from the Declaration that are relevant will be commented upon.

A collective responsibility is recognised to build a future that is based on sustainable development, which is explained in point five as consisting of

> the interdependent and mutually reinforcing pillars … economic development, social development and environmental protection … (Johannesburg Declaration of Sustainable Development, 2002).

This statement supports the premise that climate change prevention and management as aspects of environmental protection, are pertinent to sustainable development, and thus that the World Summit on Sustainable Development is relevant to this study on climate change international relations. The Declaration recognises that climate change, including more frequent natural disasters are challenges and also that developing countries and their people are more vulnerable to the adverse effects of such changes.

The Declaration reaffirms

> commitment to the principles and purposes of the UN Charter and international law as well as the strengthening of multi-lateralism. We support the leadership role of the United Nations as the most universal and representative organization in the world, which is best placed to promote sustainable development.
> (Johannesburg Declaration of Sustainable Development, 2002).

Wolfgang Sachs[4] from the Wuppertal Institute for Climate, Environment and Energy believes that the inclusion of the above paragraph can be seen as a success in the face of pressure from the USA. This will be expanded upon later in this chapter.

Events and Discussions at WSSD

Various side events and discussions took place at the WSSD; these were reported in Earth Negotiations Bulletin on the side. A selection of these events will be briefly discussed in order to demonstrate that there were events and discussions at the WSSD that were multi-participatory, and that a variety of representatives from Germany were amongst the participants. On Tuesday 27 August 2002 one of these events was hosted by the International Coalition for Sustainable Production and

Consumption, it was entitled 'The world in 2012: Towards a ten-year plan to achieve sustainable production and consumption'. Such an event is evidence of public/private debate; the Danish Minister of Environment and Energy[5] contributed to the discussion, as did representatives of the OECD, UNEP and the Federation of German Consumer Organisations. The latter of these representatives, Edda Müller held that a systematic approach is required to tackle the demand side of sustainability and that the required framework needs to be in place to facilitate a move toward sustainable production and consumption, which includes the removal of unsustainable subsidies. (International Institute for Sustainable Development, 2002, WSSD Issue 2)

Other events that relate to the subject matter of this study included one on 'The EU energy initiative for poverty eradication and sustainable development' that was hosted by the Danish Presidency of the EU. The EU energy initiative was explained by the Danish Prime Minister, Anders Fogh Rasmussen, as aiming to overcome poverty by

> promoting sustainable economic growth, further free trade and market access, and increasing development assistance to reach the 0.7% GNP target. He explained the EU energy initiative would engage in partnerships with developing countries that demonstrate good governance and political will to address poverty. (International Institute for Sustainable Development, WSSD, 2002 Issue 6)

Rasmussen also stated that nuclear power was not included in the initiative.

On Tuesday 3 September the German government presented 'German prospects for renewable energy'. Ludger Lorych on behalf of the BMU presented Germany's goals of doubling the proportion of renewable energy used by 2010 and for half of all energy used by 2050 to come from renewable sources. Lorych stated that Germany's renewable energy strategy

> is based on research, investment support and energy acts. Lorych called for holistic calculations of the economic benefits of renewable energy, and explained that the German ecological tax reform internalizes external costs into an additional tax. He presented an energy strategy for Germany that would reduce German greenhouse gas emissions by 80% through: fading out of nuclear and coal energy sources, and promoting biomass, wind and solar energy. (International Institute for Sustainable Development, 2002, WSSD Issue 8)

Franz Trieb of the German Aerospace Centre also gave a presentation, during which he detailed various technologies that utilise solar energy. Trieb stated that the costs of renewable energy technologies and fossil fuel technologies would be similar within ten years. To make renewable energy investments feasible in the short term he 'called for collaborative efforts by investors and governments' (Ibid). This presentation demonstrates that, at least some sectors of German industry are supportive of the move toward renewable energy that is strongly advocated by the German government. It is also an example of multi-participation in a discourse that aims to influence investors and governments by the force of the better argument, which is the method of decision-making advocated in Habermas' theory of discourse ethics.

Schröder's address to the Summit

Chancellor Gerhard Schröder's speech on 2 September 2002 started by highlighting the various floods and extreme storms that had recently occurred not only in Germany but also around the world. He asserted that climate change was a "bitter reality" and that "our children's future" is at stake. He appealed for the Kyoto Protocol to be ratified and

> to those industrial countries which are not acceding to the Protocol at least to make an equal contribution towards reducing greenhouse gases.
>
> The key to effective climate protection, and to successful economic development is sustainable energy supplies. (Schröder, 02-09-02)

Schröder stated that Germany had managed to reduce CO_2 emissions by 19 per cent and is on a future course of using energy more efficiently and of massively developing renewable energies. He announced that Germany would host an international conference on renewable energies, that it would join in the newly decided global energy agency network and that it would

> develop its successful cooperation in the energy sphere with the developing countries into a strategic partnership.
>
> Over the next five years Germany will provide 500 million euro to promote cooperation on renewable energies. (Schröder, 02-09-02)

The above details the majority of Schröder's speech. He goes on to talk briefly of the need for poverty alleviation and provision of clean drinking water.

Schröder's speech and especially the announcement of the hosting of an international renewable energies conference, can be seen to demonstrate Germany's continuing desire to take a lead in the international politics of climate change. It is also the case that the speech was well received in Germany (newspaper reports on Germany's efforts at the WSSD will be examined in section 6.1.2). It is worth noting that the German parliamentary elections were held in September 2002; Schröder's strong stance on climate change and relating it to the then recent floods in parts of Germany, could also be seen as playing to the domestic arena.[6] This speech may have benefited Schröder's election prospects, but it cannot be seen as purely a ploy to strengthen his domestic political reputation. As has been seen throughout this study, Germany has consistently taken a lead position with regard to climate change, regardless of the residing government's political persuasion. It is also the case that Germany under Schröder's red/green coalition government had continued this legacy, and it can be argued, had reinvigorated and intensified action in climate change politics and related policy processes.

As detailed above, Germany agreed to give 500 million euros for renewable energy projects over five years. An article entitled 'Earth Summit: After days of intense negotiations, leaders settle on a blueprint to keep the planet alive' (Independent, 03-09-02) details pledges made by world leaders at the Summit, in fact only three countries are noted as having pledged specific amounts of money, (the other two pledges were from Japan for emergency food aid for children in Southern Africa, and from Italy for debt cancellation). In a different report it is stated that Schröder promised one billion Euros 'to help developing countries gain access, to clean efficient energy. ... "... to be funded by Germany by a good E500m over the next five years, and another E500m are going to be spent for an increase in energy efficiency and energy use efficiency too" he added.' (Online.ie, 02-09-02)

Other German efforts

Germany's Environment Minister, Jürgen Trittin held a series of bilateral meetings, the first of which was with the Head of the USA delegation Paula Dobriansky (Der Spiegel Online, 02-09-02). *(Nach seiner Ankunft am Morgen will Trittin eine Reihe bilateraler Gespräche führen. Als erstes trifft er die Leiterin der US-Delegation, Paula Dobriansky.)* Other members of the German delegation also spent time networking.[7] There were numerous German attendees in Johannesburg; this in itself increases the potential to maximise influence. Karsten Sach[8] of the BMU has stated that Germany tries to keep track of and influence all negotiations in the international arena and therefore large delegations were always in attendance. However, Andreas Krämer[9] of Ecologic put forward the view that Germany had large delegations because of different viewpoints held and also because those people with differing viewpoints do not always trust each other, he did acknowledge that large delegations were also positive in that extra numbers could mean extra influence.

It is perhaps worth mentioning that Germany also supported the summit in practical terms. According to a media release from the Johannesburg World Summit Company (JOWSCO) (01-08-02), Germany contributed eight million rands to the funding of logistical requirements of the summit. Four and a half million rands went towards the Ubuntu Village in which cultural and side events were held. The German government had a stand in the Ubuntu Village where sustainability and environmental awareness raising events were held. It is also stated that the German government funded the attendance of various organisations including a number of African NGOs. These actions demonstrate that Germany enabled multi-participative discourse at the WSSD. The media release also states that the German company, Daimler Chrysler donated one thousand vehicles for transporting Heads of States and Delegations during the summit and that it was the official international transport sponsor for Jowsco. Clearly this is not done purely in the interests of philanthropy, and one can question the environmental soundness of using so many vehicles, but one could also argue vehicles would be used anyway, so why not self promote?

Numerous events were held at the German stand in the Ubuntu Village. An example of which, was entitled 'Sustainable Development and the Role of Solar Thermal Power', it was hosted by Jürgen Trittin (Minister of the Environment) and Heidemarie Wieczorek-Zeul (Minister of Cooperation and Development) on 3 September 2002. Trittin gave a speech at this event during which he admitted that he would have liked to have seen more ambitious and decisive goals agreed at the summit. He acknowledged that developing countries would need to increase their use of energy but urged them not to make the same mistakes as industrialised countries had, "Germany included". He asserted that renewable energies could be economically sound, creating jobs as has been shown in Germany. He said that by the middle of the century Germany should have expanded its market share of renewable energies to 50 per cent. With regard to assisting developing countries, Trittin stated that a partnership between Germany, the GEF, the *Kreditanstalt für Wiederaufbau* (KfW - Credit institution for reconstruction) and representatives from sixteen other countries had been formed to bring

> together interested parties and stakeholders from across the world and jointly create opportunities to use solar thermal power plants. (Trittin, 03-09-02)

The environmental impact of such a summit is a point that was not ignored. A facility was established through which attendees at the summit could offset their CO_2 usage by funding various environmental projects in Africa that would have the effect of absorbing the amounts of CO_2 released by the actions of attendees. The German government contributed to this 'Johannesburg Climate Legacy 2002' so that their trip to the summit could effectively be CO_2 neutral. It was calculated that flights and transport and energy consumption whilst in Johannesburg for one hundred and eighty German participants would produce some 1280 tonnes of CO_2, and that this could be offset by providing over 10,000 euros to a project that builds energy saving homes in South African townships.

It can be seen that a variety of ways and means were used, by a variety of German participants to influence the international political discourse of climate change.

Comparison with German/EU aims

Some aspects of German and EU influence have been highlighted in the preceding text, as have a number of instances where the text of the official WSSD documentation is reflective of Germany's ideals and policies. This section elaborates on German influence, and discusses instances where the outcomes of the summit were not as Germany would have wished.

Much of the coverage relating to the WSSD in the German press, which included a number of interviews with members of the German delegation, included the importance of climate change and its management. This subject was addressed to some degree at the summit and certainly issues regarding energy consumption and sources that have a profound effect on climate change were intensely debated. Nevertheless the relative weight that German press coverage, which presumably reflects to some degree the importance that the German populace places on this issue, was far greater than that given to this issue in the official outputs from the WSSD.

The official documentation calls for an increase in the use of energy from renewable sources and in the efficient use of energy. This topic was hotly debated. According to a report in the Earth Negotiations Bulletin (International Institute for Sustainable Development, 2002, WSSD Issue 4), many countries wanted the WSSD to adopt time-bound targets for increasing the usage of renewable energies. The common preferences were for increases of either 10 or 15 per cent by 2010. During an energy plenary some of the views reported were: the opposition of Nigeria to targets if no plan regarding funding to implement such targets were produced; Slovenia, Sweden and New Zealand wanted subsidies for non-renewable energy supplies to be stopped; and Brazil, Tuvalu, Chile and Mexico advocated the implementation of the Kyoto Protocol. It is worth noting that Germany was represented at the seventh plenary meeting in which renewable sources of energy, conservation, and energy efficiency were discussed. The only other EU Member State that is noted as being present was Denmark, which was representing the EU; Slovenia an EU candidate country was also present. The presence of Germany indicates the importance that is placed on this subject matter; it can also be argued that mere presence denotes the possibility for influence to be gained.

A great deal of press coverage was given to the differing positions of various countries on the energy issue. The consensus of opinion is that the USA and the states of the Organisation of Petroleum Exporting Countries (OPEC) blocked the inclusion of any renewable energy targets in the agreement. In the German press it was reported that there was 'bitter resistance' from the above mentioned States to the renewable energy targets that the EU and 'especially Germany' wanted i.e. 15 per cent of renewable energy to be used worldwide by 2010. Heidemarie Wieczorek-Zeul, the Minister for Cooperation and Development is cited as having criticised the USA and OPEC stating that they were "devastatingly shortsighted" and that "thinking like dinosaurs was not future competent" (Der Spiegel Online, 05-09-02)*("verheerende Kurzsichtigkeit der Opec und der USA, die in einem Dinosaurier-Denken verhaftet sind, das nicht zukunftsfähig ist").* The EU's Commissioner for the Environment, Margot Wallström is reported as having referred to the actions of the USA and the OPEC States as an "unholy alliance" (General Anzeiger, 05-09-02) *("unheilige Allianz"),* in another report Wallström is quoted as having said with regard to the agreement on energy that "This is not an ideal text for us" (Yahoo News, 02-09-02). Planet Ark reports '"The majority in the G77 has been taken hostage by the OPEC countries," one exasperated senior European delegate said' (Planet Ark, 03-09-02). The same article reports that an east African delegate 'said most developing nations wanted much greater access to clean energy sources but this was being blocked by the OPEC states.' Jürgen Trittin was more circumspect in his comments, he acknowledged that the agreement reached fell short of that aimed for by Germany and the EU, but that they had managed to have the energy problem discussed at the summit and that it 'had "not been supposed to be on the agenda two years ago". It had clearly been stated that the participating states would jointly strive for the intensified development of renewable energies,' (BBC, 03-09-02).

Doran in a report for the International Institute for Sustainable Development points out that the G77 came out against targets even though regional renewable energy targets already exist in Latin America and the Caribbean, he cites the OPEC States as being mainly responsible for this. It is also worth emphasising that countries within AOSIS, who are a part of the G77, wanted renewable energy targets. Doran apportions responsibility for blocking renewable energy targets with the OPEC States, the USA, and industry delegates. He states that

> On completing their work on the renewable energy text in the contact group, the US government delegates could be seen readily acknowledging a congratulatory 'thumbs up' from an industry colleague (Doran, 2002:12)

Wolfgang Sachs from the Wuppertal Institute for Climate, Environment and Energy, regards the failure to achieve agreement on set renewable energy targets as a "double-edged outcome".[10] The lack of fixed targets is seen as a defeat as the required formal outcome was not achieved. Sachs asserts that Germany was "pushing the line" most within the EU with regard to renewable energy targets. He argues that agreement over targets was not achieved because the G77 was divided and thus there was not a united front against the USA and Australia, he cites Japan as being between positions. Sachs believes, however, that a positive outcome was achieved because of the renewable energy initiative. Germany, he argues, in addition to its leading role with regard to its intended hosting of an international renewable energy conference, was the propelling force for the coalition of the willing, which is the subject of the next paragraph. Minister Trittin in an article for Common Ground also states that the declaration of the coalition of the willing was 'formulated at the initiative of Germany'. (Trittin, 2002)

The failure to include targets for renewable energy use in the Johannesburg agreements resulted in a large number of countries[11] forming The Johannesburg Coalition on Renewable Energy and issuing a Declaration of like-minded countries for promotion of renewable energy. The Declaration was submitted by the EU and was signed by, amongst others, all EU Member States and candidate countries and the Alliance of Small Island States. The Declaration is a commitment to promote and increase the use of renewable energy at national and global levels; it affirms the WSSD agreement but pledges to go further with regard to renewable energy; it pledges cooperation with regard to renewable energy development; it makes clear the need to reduce the burning of fossil fuels in order to combat climate change; and it commits to adopting targets for increasing the use of renewable energy. The international conference on renewable energies that Gerhard Schröder announced in Johannesburg that Germany would hold, was decided upon before his visit to the WSSD and thus before the Declaration was initiated, nevertheless it was envisaged that that the conference 'can play an important part in the implementation of this initiative' (BMU, September 2002).

The aim of Germany and many other countries to set renewable energy targets at the WSSD was thwarted. However, the General Secretary of the German Nature Protection Alliance *(Deutschen Naturschutzring – DNR)*, Helmut Röscheisen is reported as having praised the German delegation for their efforts with regard to this subject, saying that "they had fought like a lion" (Der Spiegel Online, 05-09-02) *("Sie hat gekämpft wie ein Löwe")*. Praise was also received from Germany's Organisation for the Protection of Nature *(Naturschutzbund Deutschland - NABU)* for the initiative to develop renewable energy in developing countries (Frankfurter Rundschau, 03-09-02). In the same article Helmut Röscheisen is quoted as having said that "the German's are the locomotive in the EU" and that "we are proud" (Ibid) *("Die Deutschen sind die Lokomotive in der EU", sagte …Helmut Röscheisen, am Rande des Gipfels: "Da sind wir stolz drauf")*.

Schröder's proposal for an international renewable energy conference to be held in Bonn also received 'much praise' *(viel Lob)* in Germany (Bonner General Anzeiger, 05-09-02). It was also reported that Angela Merkel a prominent CDU politician (and who in November 2005 became Chancellor) together with Bonn's mayoress, Bärbel Dieckmann, and SPD politicians Ulrich Kelber and Wolfgang Clement welcomed the move.

As previously mentioned, another issue that Germany was interested in was the formation of a World Environmental Organisation to balance the power of the WTO. This did not happen, however, agreement that would have seen WTO rules overriding global environmental treaties was avoided. It should be mentioned that Wolfgang Sachs[12] believes that Germany was hesitant in joining in discussion about the WTO and the environment.

With regard to the encouragement to ratify the Kyoto Protocol, a number of reports including that from Doran (2002) and also from Wille (Deutschland, 2002, issue 5), state that the WSSD was a turning point with regard to the Protocol. This viewpoint is due to Russia stating that they would ratify the Protocol.

A sense of how the summit was viewed in Germany can perhaps be gained from a few newspaper headlines: 'Applause for Schröder, catcalls for Powell' (Berliner Morgenpost, 05-09-02) *(Beifall für Schröder, Pfiffe für Powell)*; 'Germany failed at the world summit with proposal on energy politics' (Frankfurter Allgemeine Zeitung, 04-09-02) *(Deutschland scheitert auf dem Weltgipfel mit Vorschlägen zur Energiepolitik)*; 'World summit: No for energy change' (Der Tagesspiegel, 04-09-02) *(Weltgipfel: Nein zur Energiewende)*; 'Conference of Compromise' (Die Tageszeitung, 04-09-02) *(Gipfel der Kompromisse)*.

Sachs[13] again takes a dichotomous view with regard to the overall outcomes of the WSSD. He argues that the whole summit can be viewed "as a huge blunder when measured against the task required". However, he also argues that if seen as a struggle over whether or not multilateralism and the UN apparatus is the appropriate stage on which global concerns are managed then the summit was not a defeat. He argues that the USA and its three hundred or so delegates actively tried to undercut and dismantle the Rio process, Rio's principles especially the common and differentiated principle and the precautionary principle. The international community refuted these attempts.

Christoph Bals[14] of Germanwatch also talked of the challenge to multilaterism, he stated that world metarules were being changed, that the USA was changing towards a well-meaning imperial role.

In the opinion of Philipp Knill[15] of the BMZ, in every strategy in which you wanted to be successful the very hesitant position of the USA had to be somehow overcome; this was very difficult but in the end Germany was quite successful at the WSSD. He also stated that Schröder's speech was very welcome around the world. Although Knill expressed the above opinion, he went on to point out that Germany's influence is mainly directed through the EU.

Following on from the comment regarding the EU, it is perhaps pertinent to reiterate and expand upon some points that have been made in this and previous chapters. The EU presents a united position at meetings such as the WSSD (and COP8, which is the subject of the following section). A great deal of time and effort is spent in an ongoing participatory process that determines the EU position with regard to climate change. This participatory process continues at global conferences, as it did at the WSSD. Meetings between the EU and Member States' delegations were held every morning to ensure that a coordinated and up to date position continued throughout the conference. Meetings were also held at other times when deemed appropriate or necessary. Germany, obviously participated in these meetings. It has also been pointed out in this chapter that, at least on one occasion, German representatives were in attendance at discussions, when no other EU Member State other than the official EU representatives were reported as being in attendance. As mentioned in chapter five there is a good working relationship between members of the EU climate change working group; it is, therefore, likely that at events where German representatives were in attendance they conversed with the official EU representatives. Engaging in discourse gives the opportunity for influence.

Implications of outcomes of the WSSD for Germany and general conclusions will be examined after events at COP8 are analysed.

COP8

The eighth session of the Conference of the Parties to the UNFCCC was held in New Delhi on 23 October–1 November 2002. This was within two months of the WSSD. The general political climate that was evident at the WSSD, for example the USA's propensity to block progressive moves, was clearly going to impact on COP8. It is also the case that the climate change related aspects of the WSSD, although not a formal part of the UNFCCC process, were integral to it.

It is perhaps an obvious point as this was the eighth COP that it was part of an ongoing process, but nonetheless one that should be borne in mind. Events at COP8 and not more recent COPs are analysed because of the close proximity of the conference to the WSSD and the undoubted linkages between the two events. In 2002 it was widely perceived that the COP process was shifting from one of drafting documentation to dealing with matters of implementation, a report on this in the Earth Negotiations Bulletin states that 'the lack of buzz in the halls… seemed to confirm that the action is elsewhere' (International Institute for Sustainable Development, 2002, COP8 Issue 2:2). In addition to the COPs, meetings of the Subsidiary Body for Implementation (SBI) and the Subsidiary Body for Scientific and Technological Advice (SBSTA) take place, the seventeenth session of the subsidiary bodies met at New Delhi alongside COP8.

As mentioned previously Germany's Ministry for Cooperation and Development is keen to work with and through the GEF, this latter institution was reviewed within the SBI. In the July prior to SBI17 seven countries had submitted their reviews of the financial mechanism. These countries included Denmark on behalf of the European Community and its Member States plus Croatia, the Czech Republic, Estonia, Hungary, Latvia, Poland, Romania and Slovenia, and Samoa on behalf of the Alliance of Small Island States. In general Denmark's submission praised the GEF, though it did posit that

> the EU Member States will work in the GEF Council to further improve the timely delivery of quality-based assistance by the GEF for all types of projects. (UNFCCCa, 2002:4)

It is worth noting that Germany has one of the thirty-two seats on the GEF Council. Denmark's submission also conceded that

> The GEF should continue to improve operations at country level, i.a. through country dialogue workshops … (UNFCCC, 2002a:5)

Samoa pointed out that there had been problems with the GEF, both with the finance provided and in policies being imposed upon countries.

The draft decision at SBI17 includes the following

> the Global Environment Facility has effectively performed its role as an entity operating the financial mechanism of the Convention,
> *Welcoming* the successful and substantial third replenishment of the Global Environment Facility Trust Fund, (UNFCCC, 2002b:1)

However, it also invites the GEF

> To strengthen efforts to promote consistency of Global Environment Facility activities with national priorities and to integrate them into national planning frameworks, (UNFCCC, 2002b:2)

According to reports in the ENB and Eco there were objections to the text referring to 'successful and substantial replenishment'. Eco criticised the EU for backing the text saying that such actions were expected

> from countries that would not commit new funds, but not from the EU. (Climate Action Network, 2002, CVIII:6:3)

It is not clear whether Germany had any influence in the assessment of the GEF. However, as Germany actively works with and through the GEF, it is likely that Germany would have supported the EC's positive assessment.

Methodological issues were discussed within the SBSTA. With regard to Land Use, Land-use Change and Forestry (LULUCF) under the Clean Development Mechanism (CDM) Tuvalu on behalf of AOSIS called for social and environmental project impact assessments to be undertaken. It was agreed that a contact group would be convened to be co-chaired by Thelma Krug of Brazil and Karsten Sach[16] of Germany. The fact that a prominent member of the German Environment Ministry was made co-chair of such a group demonstrates that Germany is at the heart of negotiations. The way in which credits would be apportioned were discussed in this group, with the EU arguing for long-term crediting, Brazil wanted to limit crediting to a maximum of twenty years, Tuvalu pointed out that crediting beyond the first commitment period would be a problem, other arguments were for case by case consideration and small-scale sinks projects. It was decided that the secretariat would draw up an options paper and a workshop would be conducted in February 2003.

At the start of the high-level segment of the conference, various officials were elected. Karsten Sach was elected one of seven Vice-Presidents of COP8. As detailed in chapter five, a number of actors involved in the European climate change process have expressed the view that Karsten Sach is a respected and authoritative member of the EU working group, and that his efforts constitute a part of the German influence. Sach's election as Vice-President of COP8, and his above mentioned co-chairing of the LULUCF working group reinforces these assertions and suggests that this respect and authority is not limited to the EU level, but that it is also a factor at the truly international level.

Formal negotiations at COP8 took the form of round tables which could be seen as events at which discussions akin to the type advocated by Habermas were conducted. However, free and open speech may be where this likeness ends as it appears that rational decisions made purely on the basis of the best argument did not always result.

Round Table I – Taking Stock

Within this round table Finland made the point that the EU had made much effort to achieve definite results by 2005. The EU called for common dialogue to be entered into, an important aim of which should be the identification of a level of non-dangerous emissions concentrations. AOSIS called for immediate emissions reductions of between 50 per cent and 80 per cent.

> AOSIS and Japan said all countries will need to be involved in mitigation. Recognizing that all countries will carry different burdens with regard to mitigation. (International Institute for Sustainable Development, 2002: COP8 Issue 7:2)

The above quote is important as Germany and the EU were keen to commence talks at COP8 with regard to non-Annex 1 countries committing themselves to reductions targets in the second commitment period. It is generally portrayed that non-Annex 1 countries were against this; clearly the AOSIS countries were not. However,

> AOSIS, Mexico and Uganda noted that Annex 1 countries are not fulfilling commitments and emissions are on the rise. Considering this, Malaysia questioned how some Annex 1 countries can propose developing country emission reductions commitments. Thailand, Venezuela, Tanzania and Saudi Arabia opposed discussion of reduction commitments for developing countries. The EU underscored the need for dialogue on this matter. (International Institute for Sustainable Development, 2002, COP8 Issue 7:2)

In addition to AOSIS supporting the need for greater commitments in order to mitigate climate change it was also amongst those parties calling for the strengthening of adaptation measures.

Round Table II – Climate Change and Sustainable Development
At the opening of this session co-chair Valli Moosa of South Africa stated that

> the Delhi Declaration should draw links between COP-7, the WSSD, and COP-8. He highlighted consumption, and energy supply and access as issues where climate change and sustainable development meet. (International Institute for Sustainable Development, 2002, COP8 Issue 8:1)

Much of what has been said in the WSSD section of this chapter with regard to renewable energy and Germany's efforts in this regard are relevant here. Indeed some similarity of discussion, if not repetition between WSSD and COP8 occurred as can be seen in the report that

> Germany said the EU would build a coalition of like-minded countries willing to commit themselves to timetables and targets for increasing renewable energy use. (International Institute for Sustainable Development, 2002, COP8 Issue 8:1)

Calls for the use of renewable energies and energy efficiency were supported by Greece, Belgium and Spain, however,

> Stressing that combating poverty is the agreed priority, Kuwait said issues relating to renewable energy should not be introduced at this point. (International Institute for Sustainable Development, 2002, COP8 Issue 8:1)

An obvious point to make is that Kuwait is an oil-producing State and as such has a vested interest in keeping renewable energy and energy efficiency discussions and agreements from being introduced.

> The US said that its climate approach is grounded in sound economic policy and noted its commitment to reduce the greenhouse gas intensity of its economy by 18% over ten years. The US claimed that economic growth is the key to environmental progress. Germany responded by calling for "absolute" emissions reductions, noting that a failure to address climate change will

> result in economic harm. (International Institute for Sustainable Development, 2002, COP8 Issue 8:1)

Whilst Germany (nor any other State) has not convinced the USA of the need to ratify the Kyoto Protocol and to commit to specified targets for the absolute reduction of GHG emissions, the above report demonstrates Germany's active involvement in dialogue, through which it tries to influence opinion.

The German Environment Minister, Jürgen Trittin addressed the round table on climate change and sustainable development, he said that

> These issues are two sides of the same coin[17]. Because there is simply no way to achieve sustainability in a devastated world.
>
> In this respect, developed and developing countries have common but differentiated responsibilities:
>
> • We, the developed countries, have to rapidly re-direct our societies and economies towards clean energy, energy efficiency and more sustainable patterns.
>
> • The developing countries have to get their chance to develop themselves directly towards such a sustainable future without locking themselves in less sustainable energy forms such as fossil fuels and nuclear. (Trittin, 31-10-02)

Trittin referred to the WSSD, the need to link access to clean energy to the alleviation of poverty and to the coalition of like-minded countries who have agreed to work towards timetables and targets for renewable energy use. He also stated that

> For my government the focus on renewable energies is the key to the future. (Trittin, 31-10-02).

Trittin spoke of the commitment to increase renewable energy production within Germany to 12.5 per cent by 2010, and outlined commitments made at the WSSD to spend one billion euros in the following five years on renewable energy and energy efficiency projects in developing countries, and for Germany to host a renewable energy conference. He also

> point[ed] out that of our Kyoto target of –21% we have already achieved 19% today. With positive economic effects: Since we are now the world leader in wind energy production several ten thousand jobs have been created in this industry. And I would like to announce a decision taken only two weeks ago: Germany is

> prepared to reduce its greenhouse gas emissions by the year 2020 by 40% below 1990 levels, provided the EU reduces its emissions by 30% and other countries adopt similar ambitious targets, (Trittin, 31-10-02)

It is interesting to note how this announcement has been variously reported. According to the Earth Negotiations Bulletin (COP8 Issue 8:1) Germany said that it would reduce its GHGs by 40 per cent of 1990 levels by 2020 if all developed countries committed themselves to additional reductions. It was reported that the EU committed to reductions of approximately 30 per cent. However, a number of interviewees in Germany confirmed the figures given in Eco, the Climate Action Network's newsletter, which reported Germany's commitment to 40 per cent reductions as long as the EU commits to a 30 per cent reduction.[18] Germany's stance is reported as being 'a rare bright spot' (Climate Action Network, 2002, Vol. CVIII:1)

Trittin's speech demonstrates the leadership that Germany is taking within climate change international relations. Germany's position can also be seen as being fair or just. Whilst developing countries are called upon to take on board some responsibilities, i.e. to develop using clean technologies, funding is made available to assist in this endeavour and it is also made plain that industrialised countries have a greater responsibility to take action. This was also made clear in a press report published on the day COP8 began, which quotes Trittin as saying that

> In order to reduce the effects of climate change by 2012 (the goal date set by the Kyoto Protocol), the industrial nations will have to undertake significant further steps by protecting the environment, and developing nations will need to begin following suit with their first climate control measures. (Deutsche Welle, 23-10-02)

Round Table III – Wrap Up

In discussions on the content of the Delhi Declaration, Italy argued that action beyond 2012 (i.e. beyond the first commitment period) should be considered. Saudi Arabia wanted the Declaration to prioritise adaptation measures to the impacts of climate change and Annex 1 response measures; it made the point that the UNFCCC was not an energy convention. The G77/China opposed any text that would infer new commitments on behalf of non-Annex 1 countries. It should be noted that AOSIS is a part of the G77/China grouping; as Ott (2003) argues, the G77 largely disregarded the interests of AOSIS in favour of the demands of the OPEC States. Cuba opposed new commitments for developing countries on the basis that they had a right to development. The EU, however, pointed out

> that mitigation has proven to be a powerful force for technological change and economic development. (International Institute for Sustainable Development, 2002, COP8 Issue 8:2)

With regard to sustainable development

> The EU stressed that renewable energy exemplified the synergies between sustainable development and climate change. (International Institute for Sustainable Development, 2002, COP8 Issue 8:2)

Brazil and Austria both made statements supporting the need for renewable energy.

Delhi Ministerial Declaration on Climate Change and Sustainable Development

The Declaration includes the call by countries that have ratified the Kyoto Protocol, for those that have not, to do so. The EU was amongst those parties advocating this course of action. As has been previously stated, Germany has actively campaigned for ratification of the Protocol and would therefore, have been a staunch supporter of the EU stance.

With regard to national sustainable development strategies, the need to consider climate change when assessing the policy areas of water, energy, health, agriculture and biodiversity is identified. Requirements identified include international cooperation with regard to the development and dissemination of appropriate new technologies in key sectors of development, especially the energy sector. Improved access to environmentally sound energy is called for, as are actions

> to diversify energy supply by developing advanced cleaner, more efficient, affordable and cost-efficient energy technologies, including fossil fuel technologies and renewable energy technologies, hydro included, and their transfer to developing countries on concessional terms as mutually agreed;
>
> (l) Actions are required at all levels, with a sense of urgency, to substantially increase the global share of renewable energy sources with the objective of increasing their contribution to total energy supply, … ensuring that energy policies are supportive to developing countries' efforts to eradicate poverty;
>
> (m) Annex I Parties should further implement their commitments under the Convention, including, for Annex II Parties, those

> relating to the provision of financial resources, technology transfer and capacity-building, and demonstrate that they are taking the lead in modifying longer-term trends in anthropogenic greenhouse gas emissions, consistent with the ultimate objective of the Convention, … (UNFCCC, 2003:4).

As has been previously discussed, Germany is taking a lead in both the implementation of renewable energy technologies and in the pursuit of international agreements with regard to renewable energy and energy efficiency. It is possible, therefore, that Germany could have influenced the formation of the above text, however, there is no definite causal link.

Events and Discussions at COP8

As at the WSSD, ENB on the side reported on side events, although at COP 8 these were not as numerous as at the WSSD. German involvement in such events included: the participation of Axel Michaelowa from the Hamburg Institute of International Economics in the event on 'Market-based mechanisms for GHG mitigation: Issues and concerns'; an event on 'Poverty and climate change: Reducing the vulnerability of the poor', was a collaborative presentation that included the German Ministry for Economic Cooperation and Development; and the participation of Thomas Loster from Munich Reinsurance at both the 'Living with risk in changing climate' and the 'Climate change and the financial services industry' events. It can be seen that a variety of actors from Germany were involved in these side events; or in other words multi-participative discourse was being engaged in. Involvement does not necessarily equate to influence, but it does mean that the opportunity for such exists.

The UNFCCC secretariat held a special event that allowed Parties to discuss research issues, especially for recommendations which were put forward in the Third Assessment Report of the IPCC. These recommendations will not be gone into here. However, it is pertinent to note that the EU supported calls for research into contributions to climate change, whilst the USA and Saudi Arabia opposed these calls.

Other German efforts

On 1st November 2002 a reception was held at the German Embassy at which, Minister Trittin gave a speech entitled 'The Success Story of Climate Protection in Germany'. He commenced by stating that

> Climate protection is one of the key issues of the German Environmental Policy. Climate protection is a moral obligation to future generations and a prerequisite for long-term economic development. (Trittin, 01-11-02)

Trittin detailed policies and commitments that have already been dealt with in this study and will not therefore, be reiterated here. He then stated that

> Cooperation with developing countries is an essential element of our strategy for climate protection and sustainable development. (Trittin, 01-11-02)

Details of which were then outlined. It is clear that Germany is undertaking policies that have the aim of fulfilling the responsibilities that come with the evident importance placed upon matters of equity and justice.

Comparison with German/EU aims

The fact that the Delhi Declaration includes the call by countries that have ratified the Kyoto Protocol for those that have not, to do so, coincides with German and EU aims. It can also be seen as a success by the countries that have ratified the Protocol as the draft Delhi Declaration made no mention whatsoever of the Kyoto Protocol and this had met with the approval of the USA and Saudi Arabia. Ott argues that the

> The Delhi Declaration thus reflects the compromises required to arrive at an agreement and to prevent the meeting from complete failure. It will not go down in history as a major achievement, and immense pressure by the EU (especially Germany and France) was necessary to reach even this weak agreement. (Ott, 2003:4)

Germany and the EU wanted to start talking about the next commitment period, including the possibility of non-Annex 1 countries being included. This proposal was attacked by both Saudi Arabia and the USA, who portrayed the move as an attempt to force developing countries to undertake commitments and limit development. The EU's proposition was to start informal discussions in advance of the required formal negotiations start date of 2005, and for these informal discussions to see how the need for more extensive action to mitigate climate change can be met equitably and in a manner that enables developing countries to sustainably develop. Eco reports that the USA

> knowingly misrepresented these proposals as what they explicitly are not (Climate Action Network, 2002, CVIII:10:1)

The report goes on to state that

> Given that the Bush Administration cites the lack of developing country commitments as one of its main reasons for abandoning Kyoto, the US tactic gives cynicism a bad name. (Climate Action Network, 2002, CVIII:10:1)

This report is from a newsletter produced by a group of environmental NGOs; it could be argued that scathing remarks can be expected from such a source. However, Ott, an academic who researches climate change negotiations, expresses similar sentiments. He writes

> At COP8 the US acted with remarkably effective diplomatic skills to obstruct negotiations. In close collaboration with Saudi Arabia and other OPEC countries it managed to bring deliberations on several issues to a complete standstill or to ban any substance from decisions. (Ott, 2003:7)

The obstructiveness of the USA appears to be nothing new. Bretherton and Vogler quote an interview with a member of the European Commission's Directorate General of the Environment, in which it is stated that

> In climate, forests and biodiversity the EU is the only leader. Here the US is absent, blocking or destructive. (Bretherton & Vogler, 1999:99. Referencing Interview, DG Environment, June 1996)

Returning to Eco, it is argued that

> the President of the COP until COP9, India needs to reclaim a leadership role. It could start by taking up the EU's offer of basing long-term allocation frameworks on the principles of equity and fairness India has always called for. (Climate Action Network, 2002, CVIII:10:2)

To put this in context, India had been criticised for conducting a weak leadership. However, the point in reproducing this text is not to highlight India's role but rather that of the EU. The attempt by the EU to initiate discussions on future commitments on an equitable basis is perhaps another issue in which the EU (and Germany) are taking a lead, or at least taking a lead within industrialised nations (some developing countries have long called for equity). It appears that future negotiations will need to take issues of justice and equity into account. Ott argues that

> Evidence is growing that the intricate questions of "equity in climate policy" can no longer be ignored. There is also good reason to believe that the issues of mitigation, adequacy of commitments and climate justice will be inextricably linked in future negotiations. (Ott, 2003:9)

Whilst it can be seen that some of the EU's aims are reflected in the Delhi Declaration, not all of its aspirations were fruitful. The ENB reports that

> The EU submitted a statement of concern regarding the Declaration, calling on all countries to engage in common dialogue with a view to further action consistent with the UNFCCC's ultimate objective and based on the TAR [Third Assessment Report].[19] Noting views from across the board in support of new commitments for developing countries, he [the EU representative] stressed that the world is not "divided into two". (International Institute for Sustainable Development, 2002, COP8 Final Issue: 12)

Germany and the EU wanted dialogue to commence on the future commitment period; the fact that this was formally refused is generally deemed a failure. However, Karsten Sach of the BMU viewed events at COP8 with a little more optimism.[20] Sach's opinion is that the idea of discussing the future commitment period and to include developing countries was a difficult topic with the Kyoto Protocol not being in force, and this was capitalised upon by the USA and the OPEC States. Although, the aim of commencing official dialogue failed, nevertheless, the idea was put on the agenda and Ministers and umbrella groups talked about it. Ambassador Slade of Samoa (and Chairman of AOSIS) agreed that the time was right for dialogue. Saudi Arabia did not want to talk about it but to a certain degree they were forced to.

A point posited by Ott and in Eco reports, is that helped by the actions of the USA and the OPEC countries, a divide arose between the previously cooperative EU and developing countries. Actions intended to help overcome this divide are discussed in the next section.

Reports in German newspapers backed up the idea that the USA had been obstructive at COP8. One report is subtitled 'At the climate protection conference the USA foils a new goal' (Der Tagespiegel, 01-11-02) (*Bei der Klimaschutzkonferenz verhindern USA neue Ziele).* It was also reported (Der Tagespiegel, 01-11-02 & Berliner Morgenpost, 01-11-02) that Trittin had contradicted the USA, this made reference to the USA's claim that limiting greenhouse gas emissions is bad for the economy, and the fact that Trittin had responded by saying that Germany's economy had benefited from mitigation measures it had taken.

Implications of outcomes of WSSD and COP8 for Germany

Neither the WSSD nor COP8 produced particularly groundbreaking results and certainly none that will force Germany to adopt policies that are antithetical to their ideals. In many ways German politics will continue in the same vein as prior to the WSSD and COP8. However, German politics and policies are affected in some ways by these conferences.

Section III, point nineteen of the WSSD Plan of Implementation specifies the need for market policies and improved market signals to encourage energy systems that are compatible with sustainable development. The removal of market distortions, restructuring of taxation and the phasing out of harmful subsidies are also called for. In many ways Germany can be seen to be already embarked upon this suggested policy route. Germany already has feed-in laws that ensure renewably sourced energy is bought by the grid at a set price; one that is higher than electricity can be bought by the consumer. Germany has also made some reduction in coal subsidies. However, although the amounts of these two sectors subsidies are moving towards one another there is still a large disparity; subsidies and incentives for renewable energy are miniscule compared to subsidies given to the coal industry. If the Plan of Implementation is to be adhered to within Germany then continued movement in the distribution of subsidies would need to occur. This may give more influence to the BMU vis-à-vis the BMWA, the latter of which appear to be keener on coal subsidies than on subsidies for renewable energies. Thomas Frisch[21] of the BMWA argues that the coal subsidies make up for the difference in production costs between Germany and the rest of the world and are therefore, sustainable, as they did not lead to a price distortion. He also stated that subsidies for renewable energies had enabled the creation of jobs and now the industries had achieved a momentum, it perhaps should be seen if these jobs are sustainable without subsidies. International environmental agreements tend to favour the influence of the BMU within domestic politics as such agreements can be used as evidence of the need to act. Frisch asserted that the BMWA also generally favour international agreements because they are favourable to industry because of the worldwide approach (apart from the USA).

At COP8 the EU tried to commence dialogue regarding the future commitment period (post 2012), and to include discussion regarding some developing countries making commitments to reduce GHGs. As previously discussed, no formal discussions were agreed upon. Peter

Fischer[22] from the AA asserted that Germany wants to start a dialogue as soon as possible with regard to the future commitment period; that the EU tried to introduce such a dialogue at COP8 but that the G77 would not allow any mention of this, as they do not believe in the commitment of the industrial countries. Fischer talked of the need for more 'outreach' as a confidence building measure to overcome the lack of trust from 'the South'. He made the point that the lack of trust was not mainly directed at Germany, because although it is a large emitter of GHGs, it is one of the few countries that has anything positive to show regarding mitigation, it therefore, has some credibility in discussions calling for GHG reduction commitments. To discuss how to overcome distrust and how to reach out to the G77 countries, a conference was held in Berlin in May 2003 with the EU and accession countries. Karsten Sach[23] of the BMU also talked of the need to conduct outreach talks and workshops regarding future commitments. The planning of these outreach talks result from events at COP8. There have also been implications regarding future policies to pursue aims. Sach argues that a lesson to be learned is to carry out more outreach work before taking issues up in plenary. Responding to the failure to get agreement on future commitment talks, by determining to conduct outreach work, i.e. having conversations with numerous parties is demonstrative of the German propensity to find a discursive route to reach agreement.

The decision at COP8 to make the CDM operational could help encourage German industry to invest in developing countries, which would in turn enhance the German government policy for the private sector to be involved in cooperative developmental projects. Much of German industry has been against the introduction of the EU wide emission trading scheme, believing that they had already undertaken enough voluntary measures with regard to emissions reductions. A global system could help convince these sectors of German industry that they will not be disadvantaged.

It is likely that those people and ministries within Germany that are proactive with regard to climate change politics will carry on with their commitments and diversify their attempts to convince other countries to see things 'their way'. In Habermasian terms, they are likely to increase their discursive efforts and increase the amount of people with whom they enter into discourse. It is also likely that the BMF and BMWA will continue to try to limit the amount of actions committed to by Germany; the fact that the USA and the OPEC countries blocked any far reaching agreements at the WSSD and COP8 will continue to strengthen the argument that Germany industry may be disadvantaged if too great a lead is taken by Germany.

Conclusion

Observations regarding theory, justice considerations and German influence at the WSSD have been made in the main section of this chapter. A brief summation will be given here. Theoretically, aspects of discursive ethics in practice were observable, as were elements of justice considerations, including that of participatory justice. The discursive nature of the UNCED process of which the WSSD was a part, and the UNFCCC process of which COP8 was a part, lends itself to being viewed through a lens of discourse ethics. Much criticism has been levied at both the WSSD and COP8 for not being progressive enough. This can actually be a feature of consensual decisions arrived at through open discourse. Universalisable agreements are not likely to contain extreme positions (at either end of the scale); however, the probabilities of such agreements succeeding are greater than agreements with greater aims but that are reached through coercion.

Germany and the EU wanted targets for energy efficiency and renewable energy use to be included in the WSSD Declaration. This aim was not successful, however, following a German initiative that was channelled through the EU, the Johannesburg Coalition on Renewable Energy's Declaration of like-minded countries for promotion of renewable energy, was agreed upon. This together with the announcement by Schröder that Germany would host a conference on renewable energy demonstrates that Germany does exert influence in the international relations of climate change. The vigorous promotion of renewably sourced energy and energy efficiency by the German delegation indicates that successful domestic policies give Germany credibility in the international arena, and that there is reflexivity between domestic and global politics. Germany's success in terms of increasing renewable energy supplies and usage enabled the German delegation to credibly push for international policies to be implemented. Efforts to increase the use of renewable energies internationally and the agreed need for transferable technologies and public/private partnerships are likely to result in there being opportunities for German firms to capitalise on their expertise. The strong speech given by Schröder linking the 2002 floods in Germany (and elsewhere) to climate change and the need for renewable energy use, was positively reported in the German press. This had a positive, though probably not a decisive, effect on Schröder's election prospects in Germany. The speech was also generally well received internationally, and thus helped perpetuate the idea that Germany is a leader in climate change politics.

There are a number of instances where Germany's opinions coincided with decisions and points made. An example of which is the call for market signals for the promotion of renewable energy; Germany has already in place such signals in the form of renewable energy feed-in laws. Such issues show policy compatibility, but do not demonstrate a causal link and hence influence is not comprehensibly identifiable. It has been argued that there is reflexivity between domestic and global policies and politics. In this example domestic policies gives credibility to aims at the global level; it is also likely that decisions made at the global level to implement similar policies, will then impact back on Germany's domestic politics by reinforcing the hand of those people in government who wish to continue or expand this type of policy. This process can be explained in Habermasian terms by arguing that multi-participatory discourse allows the transfer of ideas (and where a policy has been proven to be successful, it can be seen that this is a good idea or perhaps the 'best argument') between various levels.

Germany contributed funds to the staging of the WSSD, more than half of which went toward the Ubuntu Village where side events were conducted. It also funded attendance at the Summit of number of organisations, including African NGOs. Thus, it can be seen that Germany encouraged and facilitated multi-participative discourse.

With regards to COP8 it appears that even the people within Germany that hold the most positive views of the Delhi Declaration believe that it is too weak. The failure to obtain greater progress could be argued to be a result of realist power politics employed by the USA. However, this would be a simplistic explanation for what was actually disingenuous use of dialogue on behalf of the USA and the OPEC countries. Discourse ethics as propounded by Habermas holds that all parties should have their say, each party should present their argument, listen to the views of others and come to a consensual decision on the basis of ethical and rational argument. Clearly the ethics of the USA and OPEC were at the very least, questionable. Discussing Habermas, Brulle explains

> Validation requires an open speech community in which the unforced force of better argument prevails. (Brulle, 2002:4)

Clearly the tactics used by the USA and OPEC did not meet this criteria.

Within the institutionalist framework of the UNFCCC COP process, consensus is required. This often leads to only incremental progress but it also prevents traditional power politics being used to completely dominate less powerful parties. As previously discussed, the Delhi Declaration was not allowed to progress in its draft form, which was approved by the USA and OPEC. The actual Declaration was a little more stringent than the draft form; it included the call by those countries that have ratified the Kyoto Protocol for those that have not ratified it, to do so. As stated above, Ott argues that this was largely the result of pressure from

Germany, France and the EU. In his opinion this saved COP8 'from complete failure'. Theoretically, this can be seen as indicative of discursive ethics working to some degree, even if only in a weak form. In terms of influential achievement, it can be seen that Germany did have an effect, this is not definitively quantifiable, but the fact that Germany had an influence, is observable.

A breakdown of issues where Germany influenced the international relations of climate change at the WSSD and COP8, and instances where there is a probability that some influence occurred can be found in table 6.1.

Influence that Germany achieves in the international relations of climate change has, so far in this study, been examined in relation to the EU and the global conferences of the WSSD and COP8. Other ways in which Germany pursues and achieves influence in climate change international relations will be examined in the next chapter.

Table. 6.1 German influence and probable influence at the WSSD and COP8

Instances where German influence can be definitely identified.

1. Johannesburg Renewable Energy Coalition (JREC) was a German initiative that was channelled through the EU.
2. International Conference on Renewable Energies to be hosted by Germany announced at the WSSD. This coincided with the push for renewable energy targets and the formation of the JREC, providing impetus and a forum where targets can be discussed.

Instances where comparison of German policies and aims with outcomes of the WSSD and COP8 indicate some influence may have occurred, but no definite causal link can be proven.

1. The reaffirmation of the Rio principles.
2. The call for states that have not already ratified the Kyoto Protocol, to do so.

Agreed need:

3. To increase use of renewable energy;
4. To increase energy efficiency;
5. For financial and technical assistance, in developing countries;
6. For capacity-building measures in developing countries, and for international cooperation to assist in this aim;
7. For public/private partnerships;
8. For transportation policies to be implemented that minimise GHG emissions;
9. For the promotion of sustainable development at national level;
10. For the strengthening of UN bodies and for increased cooperation between them, i.e. between the GEF and the WTO;
11. For environmental assessments.

7

ACTIONS OUTSIDE OF WSSD AND COP8

The previous chapter analysed events at the WSSD and COP8 in relation to Germany's aims. A great deal of effort was put into these conferences, but it should be remembered that climate change politics is not limited to such high profile events. It could be argued that the 'real action' takes place in the more routine processes that take place on a continual basis away from the limelight of large international gatherings. This chapter will discuss these less publicised activities that are part of the ongoing process of Germany's international relations of climate change. The purpose of the first section of this chapter is to highlight the variety of activities that are pursued, thus the nature of this chapter will be very different from previous chapters in that in depth analysis of events will not be undertaken. As just mentioned, the first section of this chapter will look at various pathways that Germany takes in pursuing its aims. These various sub-sections overlap to some degree, for example, some bilateral projects are reported through the UNFCCC national communications process. Nevertheless, there is enough of a distinction between actions to be able to divide them into sub-sections. The second section of this chapter will focus on the low-lying island States of the Pacific. Relations that Germany has with the islands will be looked at, as will the effects that Germany's international relations of climate change have on the islands. As explained in the introduction, the Pacific Island States are particularly vulnerable to the effects of climate change and it is for this reason that it is appropriate to focus on these States in order to ascertain the effectiveness or otherwise of Germany's actions in the international relations of climate change.

Ways in which Germany conducts climate change international relations

As mentioned above, this section looks at a variety of ways through which Germany conducts its climate change policies. The UNFCCC process will be looked at, as will Germany's interactions with and through the GEF. Other aspects of Germany's multilateral relations will also be considered, as will bilateral relations.

Germany and the UNFCCC

The UNFCCC secretariat is based in Bonn. Whilst the secretariat is a United Nations body and as such is independent of any one nation, the fact that it is sited in Bonn may provide Germany with a subtle form of influence. Initially, Germany offered to host the secretariat; this in itself does not confer any extra degree of influence, it does however, indicate that Germany was particularly keen on the success of the UNFCCC process. Until 1999 Bonn was the site of the German government and the mere fact of the proximity of various UNFCCC and German governmental officials would perhaps have made for ease of communication and the propensity to build upon working relationships. Whilst the German parliament now resides in Berlin, ministries do have offices in both Berlin and Bonn, it is possible therefore, that some ease of communication still remains. It is worth noting however, that the climate change staff of the BMU are based in Berlin. The professional staff of the UNFCCC secretariat are of various different nationalities, but as perhaps would be expected, Germany is highly represented in terms of the nationality of the administration staff. The national composition of staff within the secretariat is unlikely to furnish Germany with any additional avenues of influence. The opinions of the secretariat's professional staff vary with regards to the positioning of the UNFCCC secretariat within Germany being influential or not. Only three members of staff expressed an opinion on this subject, however, differences of opinion emerged. One representative expressed the belief that this did not give Germany more influence; but then said it was possible that at a high level there could be more influence, but possibly not. Other representatives believed that the location of the secretariat does allow Germany particular influence. However, it should be noted that the examples given of extra influence can mainly be attributed to extra effort that Germany makes, rather than due merely to the fact that the secretariat is sited in Bonn. One factor is that Germany as the host nation contributes greatly to finances, particularly in the form of the Bonn fund. This finances the participation of non-Annex 1 countries at meetings in Bonn. This shows that Germany is supporting a fully participative process. Clearly for all effected parties to be able to participate in discussions, as required by Habermas' discourse ethics, all parties need to be present and Germany is facilitating the attendance of the least developed countries. In addition to financing participation at meetings, the Bonn fund contributes to the funding of meetings that occur in Bonn that have been mandated by a Conference of the Parties or by the Subsidiary bodies (for example a modelling workshop). Another factor of potential additional influence is that when meetings are held in Bonn, there is full access to experts within Germany. Some countries may send only their Environment Minister, whereas German representatives can include policy-makers, meteorologists, etc. Another point made was that the UNFCCC secretariat has access to research institutions throughout Germany, and can attend conferences when invited at a variety of places within Germany as it is inexpensive to attend. Even if the conference is in German, there is a German national of the professional staff who can attend. The point was also made that Germany was interested in the UNFCCC process and was always present at discussions, not all countries, even those that can afford to attend, do. It was also stated that when the EU speaks at meetings, German representatives were always there and able to advise. The fact was also mentioned that as the secretariat is in Germany there is interaction between staff members and local people and sensitisation to local issues. This can mean that reports include issues that are important within Germany, for example the floods in Germany were mentioned. It was stated that the word 'Germany' was mentioned just that little bit more frequently in a variety of circumstances than perhaps would otherwise be the case. This is perhaps indicative of discourse being important, not only in terms of discourse ethics at formal meetings but in the general discourse employed in everyday life.

As mentioned in the previous chapter Karsten Sach of the BMU was made co-chair of a LULUCF contact group, although this appointment was made during COP8 this position and the work of the group was an ongoing project within the UNFCCC process. Co-chairship of this group is indicative of involvement in the ongoing processes of the UNFCCC; involvement gives opportunity for influence.

As a part of the UNFCCC process, countries submit national communications, which are then assessed and reviewed by the UNFCCC. At COP5 it was recognised that least developed nations require some assistance to be able to complete national communications. This resulted in the establishment of a consultative group of experts to assist those countries that need help. Experts on inventory, mitigation, adaptation, capacity-building, etc. have conducted regional workshops. These were funded by money given by the USA, Switzerland and the Netherlands. The workshop held in Bonn was funded by Germany through the Bonn fund. In addition to this contribution Germany provides one of the experts, this consultant receives payment from the German government to work for the UNFCCC. Karsten Sach[1] of the BMU made the point that Germany and the Netherlands have always funded two experts from 'the North' to help developing countries complete their national communications, and that this arrangement also allows for the developing countries to explain to the experts what their priorities are and how Germany and the developed countries could take into account their climate change concerns. This facilitates a substantial debate between Germany and individual developing countries with the aim of determining what that particular developing country could do and how Germany could help them in their endeavour. It is worth mentioning that the consultative group of experts is made up of representatives from different countries from different regions, there are five experts from the African region, five from Asia, five from Latin America and the Caribbean, six from Annex 1 countries (this includes the German expert), and two from organisations. The point of mentioning this is to show that not all Annex 1 countries provide this type of support.

An example of Germany continuing to work through the UNFCCC process is that it hosted a workshop, in Berlin in December 2002, for the Annex 1 expert group on policies and measures in industry. Another workshop was held in Leipzig in March 2003 for middle and east European countries to discuss Joint Implementation and Emissions Trading. Whilst this was not strictly speaking within the UNFCCC process, its subject matter deals with the Kyoto Protocol implementation strategies and for that reason it has been included in this section. The workshop was organised by the BMU and the Wuppertal Institute for Climate, Environment and Energy.

Influence through the UNFCCC process is not quantifiable; however, it is clear that Germany is an active participant. It is also clear that Germany supports participation in the process by countries that would otherwise find it difficult, if not impossible to do so. This can be understood as influence in that these actions contribute towards the potential for the UNFCCC process to succeed. It is also reflective of German national values and practices of multi-participation.

Germany and the Global Environment Facility

The GEF is linked to the UNFCCC in that the former is a funding facility through which UNFCCC implementation projects are funded. The GEF is, however, separate from the UNFCCC. The GEF also deals with the funding of a variety of environmental projects including those that address issues concerned with biodiversity, international waters and ozone depletion.

As mentioned in the previous chapter Germany is keen on working through the GEF, and as such takes an active role in its running. The GEF Council consists of thirty-two members who represent set constituencies.[2] There are sixteen members from developing countries, fourteen from developed countries and two from countries with economies in transition. Germany has a seat on the Council, its constituency is Germany. Some representatives cover the interest of a number of countries in their vicinity. Clearly having a member on the GEF Council confers the potential for influence to be achieved.

As previously stated, Germany is the third largest contributor to the GEF funds,[3] behind the USA and Japan. Although council members come from a variety of countries with widely varying economies and contributions to the GEF and all decisions are consensual, it is likely that those who contribute large amounts to the fund hold some degree of influence over how such funds are spent. It is the opinion of Philipp Knill[4] of the BMZ, (within which he has responsibility for climate change and the GEF) that Germany does have a say in how the GEF funds are spent and that it has a great deal of influence regarding strategy and policy approval. Knill cited the fact that Germany has a seat on the Council and that the Council is most important within the GEF regarding strategy and day-to-day business. The Council is also responsible for project approvals, although Knill stated that Germany did not try to influence the Council to concentrate on particular countries. However, the BMZ links bilateral aid to multilateral aid with the aim of having greater impact and in a few countries the BMZ has co-financing with GEF projects; examples of which include a solar thermal power plant in India and biodiversity projects in Ecuador. With regard to the GEF it was also stated that Germany is active in commenting on projects and strategy and that at the suggestion of Germany and Canada the monitoring and evolution unit was being reorganised to become more independent and to use lessons learnt for future projects. In chapter six of this study it was reported that the WSSD Plan of Implementation encourages the GEF to

> leverage additional funds from key public and private organizations, improve the management of funds through more speedy and streamlined procedures and simplify its project cycle. (WSSD Plan of Implementation, 2002).

The reorganisation or the monitoring and evolution unit could be an example of Germany, together with Canada influencing the GEF to streamline procedures and simplify its project cycle.

It is perhaps worth mentioning that GEF projects that fall under the climate change category include those that address: energy efficiency;

renewable energy; low GHG technology; sustainable transport; and targeted and streamlined capacity building. A list of funding for climate change projects in a variety of countries can be found on the GEF website.[5] Influence that Germany may or may not have had in the allocation of these funds is not apparent.

Germany's influence within the GEF is not quantifiable. However, Germany is active within the GEF, it is a major financial contributor to the GEF, and it is keen to work through the GEF. It is the belief of the BMZ that Germany has influence with the GEF, and an example of streamlining of a GEF unit at the suggestion of Germany and Canada has been given. It can therefore, be surmised that Germany does influence the GEF process.

Multilaterally

The international relations considered thus far in this study have largely been of the multilateral variety. This sub-section will point to some other facets that constitute Germany's multilateral international relations that are linked to climate change.

One way by which Germany engages in discourse with other countries is through the use of workshops. The BMU has funded capacity building workshops, often through the auspices of UNEP; attendees have included African and AOSIS countries. These workshops concentrate on different aspects of climate change management, for example on how to deal with the CDM. One particular workshop was held for African ministers in order for them to coordinate their position and to strengthen their negotiating capacity. Another example is the financing of a network in India that holds workshops for developing country and NGO representatives. Workshops for developing countries to increase understanding of climate change have been held by the Ministry for Education and Research (BMBF).

In chapter six, in the discussion on COP8, the lack of trust that some developing countries have in the developed world with regards to the latter's commitment to reducing GHGs was raised. It was asserted that Peter Fischer of the AA had expressed the need for confidence building measures to be introduced with the aim of overcoming this problem. An initial step that the AA took toward such confidence building measures being introduced was to hold a meeting in Berlin in May 2003 for foreign office environmental officers of EU Member States and accession countries to discuss what needs to be done to overcome this lack of trust. In order to do this, the viewpoints of the developing countries were discussed, as was the best way to formulate a communication strategy that takes into account the best way to talk with the developing countries and with whom to talk. A German proposal for the establishment of information networks was for foreign embassies to be better and systematically utilised for the dissemination of climate change information. This will be discussed in a little more detail in the bilateral relations sub-section.

As discussed in chapter six, the Johannesburg Renewable Energy Coalition (JREC) that was formed at the WSSD although presented by the EU was a German initiative. This initiative is a prime example of Germany working with others in a discursive manner in order to try and realise its aims. The JREC secretariat is based in Brussels within the European Commission. Nevertheless, Germany is still playing an instrumental part in its continuance. A JREC roadmap was agreed in June 2003, it includes four regional meetings, one of which, the European Conference for Renewable Energy – Intelligent Policy Options was held in Berlin on 19–21 January 2004. As pointed out in chapter six, decided upon prior to, but announced at the WSSD, was the German intention to host an International Conference for Renewable Energies. The conference, which was held in Bonn on 1–4 June 2004 provided a forum where the aims of the JREC, to work towards agreed and binding renewable energy targets along with timelines within which these targets should be reached, were discussed. This conference is cited as a key international conference in the JREC roadmap. The conclusions of the European Conference for Renewable Energy provided inputs for the International Conference for Renewable Energies in Bonn. Information available prior to this latter conference stated that it was to be

> convened as an "extended government conference".
>
> Among those invited and expected to attend are:
>
> • ministers from all UN member states responsible for environment/energy and development cooperation
>
> • government representatives at implementation level with the above portfolios
>
> • international and multilateral organisations, international financial institutions
>
> • private sector and civil society representatives and NGOs, including the principal renewable energy networks
>
> About 900 official delegates are expected.
>
> Parallel activities peripheral to the conference will address in particular the business and scientific communities and representatives of other national and international interest groups.

> In addition, the German federal parliament (Bundestag) plans an international forum of parliamentary members.
> (www.renewables2004.de/en/2004/particpants.asp?page=print October 2003)

The conference was actually attended by representatives from one hundred and fifty four countries; the total number of delegates (not just government representatives) exceeded three thousand.

As already indicated it is not the intention in this section to examine in minutiae the activities that are outlined. The above is reproduced in order to demonstrate the multiple avenues and sections of the global community that the German government is including in its discursive endeavour. A further example of the multi-participative nature of German climate change related international relations is that the Wuppertal Institute for Climate, Environment and Energy[6] was involved in the conference's preparatory process and is also involved in assessing its results. Although media coverage of the conference was not as prevalent as that for the WSSD, it nevertheless widened the audience of the renewable energy discourse. This will have raised awareness of the subject matter and also the fact that Germany hosted the conference, this will have added to the generally held opinion that Germany is a major player in this issue area. Whilst the audience of the media output did not have an opportunity to engage in the discourse as would be a requirement of Habermas' discourse ethics ideal, it is possible that any recipient of the message for the need for increased renewable energies could in the future participate in the debate.

Other conferences that Germany has held in the past that can be related to climate change include: The Environment, Sustainable Development and Trade, held in Berlin on 20–22 March 2001, this conference was organised by the BMU and UNEP; and the International Conference on Freshwater that was held in Bonn on 3–7 December 2001.

It is also worth bearing in mind that there are a number of initiatives that are not directly run by the German government. An example of this type of programme is a project entitled 'South-North Dialogue – Equity in the Greenhouse'. The dialogue was initiated by the Energy and Development Research Centre of the University of Cape Town and the Wuppertal Institute for Climate, Environment and Energy, and the project was commissioned, and supported, by the BMZ and the GTZ. The first phase of the project has involved dialogue among fourteen research institutions from around the world and has been inclusive of developed and least developed countries and those that are most vulnerable to climate change.

> The purpose of the "South-North Dialogue" is to discuss building blocks of a future international framework to combat climate change in a *participatory* manner. This comprises issues related to mitigation as well as adaptation and will be based on the underlying principles of "equity", "adequacy" and "development". (Wuppertal Institute, 2003a: 3)

The BMZ has established an advisory board that includes representatives of governments, research institutes and NGOs. This project is a multi-participatory and discourse based programme aiming to determine a consensually arrived at equitable solution to the problems of climate change. It appears to have many of the attributes that Habermas' discourse ethics, demands. It is also demonstrative of the apparent German propensity to arrive at solutions to problems through fair and just means with the aim of arriving at just policies. Not only do these attributes apply to the project, they also apply to the findings. The final report of the first phase of the project 'South-North Dialogue on Equity in the Greenhouse: A proposal for an adequate and equitable global climate agreement' details the requirement for the:

> *Capacity to engage politically:* An equitable process (i.e.) procedural equity is a precondition for an equitable outcome. (Ott *et al,* 2004:8)

The second phase of the project was due to commence in late 2004 with the aim of expanding the dialogue to the political level.

This sub-section has discussed a variety of German initiated multilateral relations. It is of course, also the case that Germany participates in multilateral events organised by non-German parties. It is not possible within the constraints of this study to be able to investigate all international relations in which Germany is involved. However, it can be assumed that German aims at other multilateral events will be consistent with those outlined in this study and that the German modus operandi will be to pursue a consensual agreement on how to proceed through multi-participatory discourse.

Bilaterally

Multilateral relations are probably the most important element of the various means of dealing with climate change and these have been discussed at length in this study. Although multilateralism is extremely important it is at national and local levels that actual implementation occurs, chapter three detailed German efforts to manage climate change. In addition to local and national governance and multilateral relations, bilateral relations occur. This sub-section will look at the relations that Germany conducts bilaterally. Some bilateral relations are conducted on a formal intergovernmental basis, whereas others are carried out through more informal working relationships between people within governmental departments. Bilateral relations include political meetings, diplomatic efforts and cooperative implementation strategies. It is only possible to give an overview of these multifarious connections. To give an indication of the type and variety of contacts a selection of the more prominent bilateral relations will be outlined.

Mentioned previously was the German proposal to utilise their embassies in the task of overcoming developing countries lack of trust in 'the North'. As posited by Peter Fischer[7] foreign ministries have a very valuable tool at their disposal in the form of its embassies. The embassies have an advantage in that they are permanently sited in a given country and are able to establish a dialogue with the relevant sections of the host country government, they can also speak to civil society, the media, business associations, companies and environmental associations. This desire to have a dialogue with diverse sections of society can be seen as reflective of the belief in discourse ethics as espoused by Habermas; it can of course be seen purely as using any and all means necessary to try and get one's message across. A concrete example of the use of German embassies is that they have been given information on how flexible mechanisms work; shortly after this information was disseminated the German embassy in Bangkok arranged a seminar with German companies in Thailand and representatives of the Thai Environment Ministry. This raised awareness of the CDM in German companies and of the German companies in the Thai Environment Ministry, both parties found this useful and agreed to review the situation within a specified timeframe. Karsten Sach[8] made the point that utilising the embassies to engage in a continual dialogue with developing countries would enable Germany to take the view of the developing countries on board at an earlier stage so that as far as is possible clashes can be avoided at international meetings. Sach also stated that this type of dialogue would also enable Germany to get the message across that the position taken by Germany and the EU was not hostile to the developing countries, but that dialogue was needed on how best to move forward. As mentioned previously, Germany is also trying to persuade other countries within the European Union to use their foreign embassies in a similar manner; this shows that some actions that can be broadly defined as bilateral also entail elements of multilateralism. The German idea is that where a particular country's embassy has either particularly good relations with the host country or where an embassy has particular expertise in climate change issues, then that embassy should be used to spearhead the campaign.

A continuing theme in this study has been of Germany working with and through the EU. As well as working along side all Member States at a variety of meetings, it is also the case that there are ongoing contacts between individual Member States and of course the European Commission. As was mentioned in chapter five, members of the EU working party on international environmental issues – climate change have built up a close working relationship and it is the case that telephone calls and e-mails are exchanged between members and their teams as the need arises. It has been asserted by representatives[9] of both Germany and the UK that these two countries have a particularly good working relationship.

A bilateral contact outside of Europe but also one where viewpoints are similar can be found in Japan. Examples of this relationship include a 'Policy Dialogue between Japan and Germany for Facilitating Co-ordinated Measures to Address Global Warming'. The dialogue was initiated by the Japanese Institute for Global Environmental Strategies and the Wuppertal Institute for Climate, Environment and Energy. The first phase of the dialogue included a 'Japanese-German workshop held in Tokyo in March 2000 and an informal meeting of business stakeholders with the organisers and representatives of both governments' (Ott & Takeuchi, 2000:V). Germany and Japan tend to hold similar positions with regard to the UNFCCC process and following an approach to the AA from Japan,[10] bilateral meetings have and will be held prior to Subsidiary Body meetings, Conferences of the Parties and major conferences, in order to try to coordinate positions.

A case where official government positions diverge is between Germany and the USA. Contacts between the two countries exist via a number of routes, for example: between the governments; with some individual States, for example California,[11] who are interested in more progressive policies; between departments; and through the EU. It is the German policy[12] to keep dialogue open with the USA in the hope that the gap between outlooks does not become wider and that in the long term the USA may change its policies so that they become closer to those of Germany. It is the case that some individual States and companies within the USA are far more progressive with regard to climate change policies than are their government. With regard to relations with these individual States, it is the German policy to foster dialogue and if and where possible, practical cooperation.

Diplomatic efforts are also an ongoing process between Germany and Russia. It has been a fervent desire of Germany (and many others) for Russia to ratify the Kyoto Protocol. As explained in chapter three, when Russia completed this procedure, the required ratification by 55 parties to the convention making up at least 55 per cent of 1990 emissions of Annex 1 parties (industrialised countries) had occurred and the Protocol entered into force in February 2005. As already stated, up until Russia ratified the Protocol diplomatic pressure for them to do so was ongoing and as such not all avenues and occasions of contact can be detailed. A few instances of contact will be mentioned as a sample of such efforts. Much of this diplomacy is likely to take place invisibly. An instance where this type of contact was revealed was toward the end of a meeting with a German government employee when a brief telephone call was taken, the comment was made that it was a colleague in Moscow who was trying to encourage ratification. A more concrete example of contact occurred in early February 2003 when Chancellor Schröder wrote to President Putin urging Russia to ratify the Kyoto Protocol.[13] The facts that it would be in their joint interests for Russia to ratify the Protocol, and that joint implementation projects could be put into practice were included in this appeal. Another specific example of dialogue was detailed in a BMU press release in October 2003 that reported on environment talks between Germany and Russia. It was reported that Simone Probst, Parliamentary State Secretary to the BMU called for Russia to ratify the Kyoto Protocol. It was also stated that during the two day bilateral event cooperative projects totalling in excess of one million euros were adopted. It was stated that

> In addition to climate protection, these projects for the 2004-2005 period comprise sustainable development, handling genetically modified products and international cooperation in environment policy.
>
> The Federal Environment Ministry has already supported more than 60 bilateral projects in the context of German-Russian environmental cooperation and provided more than Euro 7.5 million. (BMU, 29-10-03)

The report also states that Germany and Russia were preparing joint climate protection projects that are scheduled to begin in 2008.

China is another large country with which Germany has bilateral meetings. The German-Chinese Environment Conference was held in Beijing on 12–13 December 2000. A report in the BMU's publication 'Common Ground', states that the conference resulted from an initiative put forward by Chancellor Schröder and the Chinese Prime Minister, Zhu Rongji. It is stated that

> The event was attended by no less than 1,110 businesspersons, scientists, officials and representatives of NGOs.
>
> Fifty six German and 350 Chinese companies participated in the 'contact bourse' accompanying the conference. Some 130 German-

> Chinese ET (environmental technologies) business relationships were forged at the bourse, reported the Süddeutsche Zeitung. …
>
> A key focus of German-Chinese efforts is the revamping of China's energy supply system. This entails the modernizing of the country's power plants and the fostering of the use in China of regenerative and other environment-friendly forms of fuel. Germany is already supplying DM165 million to promote the installing of photovoltaic facilities and other solar-based equipment on China's roofs. Germany is now adding a further DM 20 million to this. Another DM10 million will go to promote the environmentally-friendly development of China's cities. (BMU, 2001: 11)

It was agreed at the conference to hold regular environmental forums to discuss possible solutions to China's environmental problems. The first such forum was held in Beijing in December 2003, it was organised by the BMU, the German Trade and Industry's Asia Pacific Committee, the Chinese State Environmental Protection Administration and the China Federation of Industrial Economics. The forum focused on improving energy efficiency and the use of renewable energies in China. At the opening of the forum Margarete Wolf, German Parliamentary State Secretary stated that she was

> confident that this forum will contribute to an experience exchange in the fields of environment and energy, in preparation for the International Conference for Renewable Energies in June 2004. (BMU, 19-12-03)

These conferences could be viewed as glorified trade missions. However, whilst it is obviously the case that German businesses are benefiting, China is also benefiting as it is gaining access to advanced technologies. These are enabling China to develop and to increase the population's access to energy supplies, without going down the purely damaging industrial route that the developed countries have done. This also has worldwide benefits as the damaging effects of climate change caused by unfettered expansion of traditional forms of energy production, are minimised. These bilateral meetings can also be seen as examples of the use of multi-participatory discourse in an effort to reach further mutual understanding; in other words, in order to convince China that it can develop in a sustainable manner, or at least to do so in a less environmentally damaging manner than has been the norm and also to help it achieve such aims.

Another example of formalised bilateral contacts that have not been initiated by the German government but in which they are involved is the Indo-German Forum on International Environmental Governance. This is co-ordinated by the Global Governance Project, which is a joint research programme of the Potsdam Institute for Climate Impact Research (PIK), the Free University, Berlin and Oldenburg University. PIK is funded from a variety of sources including the German Federal Republic and the State of Brandenburg. The Indo-German Forum is financially supported by amongst others; PIK and the KfW. The Forum facilitates dialogue between academics, individuals and institutions that work in the field of environmental governance. An inaugural conference was held in Potsdam on 25–27 September 2002; participants at the conference included representatives of both the Indian and German governments. Representatives from the BMBF, BMU, BMZ, and the Bundestag attended the conference. In addition to this Indian participants were invited to the German Bundestag and met with officials from the AA and the GTZ.

Mentioned in chapter four was the fact that the BMZ supports the introduction of improved energy efficiency measures in the developing world. Also mentioned at various points throughout this study has been the Kyoto Protocol's CDM which aims to encourage investment in 'clean development' in developing countries. A BMZ climate and energy paper states that

> With assistance from the BMZ, KfW is currently devising a fund arrangement for making use of the project-tied flexible Kyoto mechanisms (CDM and JI[14]) and financing projects in developing and transition countries that meet climate protection standards. With new finance instruments, additional projects can be developed … Particularly when planning financial cooperation energy and transport projects, the possibility of finance under the CDM rules should be examined.
> (www.bmz.de/en/topics/umwelt/umwelt5.pdf February 2004)

The paper goes on to detail four projects. A project to tap new sources of power with wind farms in Egypt and one that promotes the financing of renewable energies in India are attributed to the KfW. Another project is a public-private partnership in South Africa to install 27,000 solar home systems. The remaining project is said to be GTZ assisted and is aimed at saving energy to boost competitiveness in India.

The BMZ is in charge of official international development cooperation; they have established a list of cooperation countries with which to work, this list is sub-divided into priority partner countries, partner countries and potential cooperation countries.

> The selection of priority partner countries and partner countries is based upon the *need for co-operation* in the context of our economic, social, ecological and political development targets and interests. (BMZ, 2002b:3)

Of the countries mentioned thus far in the bilateral sub-section, those that are not included in the BMZ list are Japan, the USA and Russia and these are not classified as developing countries and thus not in need of development assistance. It is pertinent to note that none of the small island developing States (SIDS) are named as cooperation countries. This fact will be discussed later in the chapter.

As explained in chapter three the GTZ is a government owned corporation for international cooperation. The BMZ and the GTZ work closely together with regard to international development projects. The GTZ has a Climate Protection Programme, through which a number of projects have been conducted with developing countries within Africa, Asia and Latin America.[15] Again none of the countries involved are SIDS. It is the GTZs remit to conduct development assistance and as such climate change considerations are an additional aspect of projects; the only purely climate change projects that are carried out are GHG inventories. However,

> The overriding principle of our climate change activities is mainstreaming into German development cooperation. Therefore, we will only be active in countries and projects which allow a close linkage between climate change and ongoing or planned development cooperation projects. (Liptow, 2003)

With regards to adaptation work the GTZ focuses on Africa. Philipp Knill[16] of the BMZ confirmed the German aim to implement adaptation measures. He stated that this is not always straightforward as not all countries are interested, however, Germany already has bilateral relations in Africa and it has the leverage to encourage adaptation measures.

It is the view of a representative of the European Commission's DG Environment[17] that if Germany has a weakness with regards to international relations of climate change it may be in terms of its bilateral contacts outside of the EU. Historically the UK has stronger connections for example to the USA, China, and Australia. However, it was recognised that Germany is taking action to overcome this historical weakness. Germany certainly works mainly within a multilateral context; however, it appears from the evidence provided within this section that Germany does have numerous bilateral contacts and is also active in strengthening and building upon them.

Summary

As has been seen in the content of this study most of Germany's international relations of climate change are multilateral. The EU is worked with and through, as are the UN and related processes and conferences. In addition to these mainstay activities, some additional multilateral communications occur. This preponderance of multilateral relations could be partially due to the transnational nature of climate change and hence the need for multinational governance and therefore the requirement for multilateral relations. Nevertheless, bilateral contacts do exist and are being built upon. The expansion of bilateral relations can be seen as an expansion of the required multifarious responses to the effects of climate change. Another aspect of the German response to climate change and environmental issues is that it contributes to the funding of the Earth Negotiations Bulletin (ENB). The International Institute for Sustainable Development publishes the ENB, which reports on environmental conferences. It has been cited on numerous occasions within this study. By contributing to the funding of this publication over which it has no editorial control,[18] the German government can be seen to be aiding the dissemination of sustainable development discourse.

A recurring theme in this study is that Germany conducts its climate change international relations through the use of discourse. One of the questions that was asked of a number of interviewees was whether or not this premise was correct, as opposed to the USA who tend to engage more in power politics. This was always answered in the affirmative and often answered with a referral to Germany's history, for example: "with our history we are always in favour of international regimes, with our history we can be successful only within international law and international regimes". It was also stated that "we believe in multilateral systems so all can have a say for example, the USA, Germany and Samoa". It was also posited that multilateral systems allow for coalitions of the willing to be formed. The inference here is that such coalitions can stand up to the perhaps otherwise unassailable power of individually mighty States, such as the USA.

This first section of this chapter has looked at numerous ways through which Germany conducts its climate change international relations. The next section will examine whether Germany's efforts have had an effect on the low-lying island States of the Pacific.

How does Germany's influence in international relations affect countries vulnerable to climaye change? Analysis of any positive (and negative) effects on Pacific Island countries.

As explained in the introduction to this study, low-lying island States are amongst the first and worst to be detrimentally affected by climate change. Some Pacific Islands are already suffering the effects of climate change; and yet they have contributed very little, if at all to the causes of climate change. These micro States also tend to be economically vulnerable and so implementing adaptation measures is not easily achievable. As Germany aims to influence the international relations of climate change so that adequate mitigation and adaptation measures are implemented, it is reasonable to assess Germany's efforts in relation to any effects that may be experienced by these particularly vulnerable States.

Before assessing the possible effects of Germany's efforts, a little background information on the Pacific Islands will be given. Many of the Pacific Island States consist of numerous small low-lying islands and/or atolls. They are, therefore, particularly vulnerable to sea level rise and to storm surges, these are both phenomena that are associated with climate change. It is also the case however, that even on larger higher islands it is generally the low-lying land along the coastline that is most habitable and inhabited. Therefore even larger islands, such as Fiji,[19] are vulnerable to the threats caused by climate change. Rising sea levels are already causing salt-water intrusion of agricultural land on a number of islands including Kiribati, the Republic of the Marshall Islands (RMI) and Tuvalu;[20] these island States are amongst those that could completely disappear. Salt-water intrusion has implications regarding access to fresh food and to fresh water supplies. Neither Kiribati, the RMI nor Tuvalu have any rivers, though Kiribati and the RMI do have some freshwater lens'. Most of the freshwater needs of the low-lying Pacific Islands are met by the catchment and storage of rainwater. Climate change is also affecting patterns of rainfall; the Federated States of Micronesia, Fiji, Papua New Guinea and the RMI have experienced drought conditions. Climate change is also expected to cause the increase in frequency and intensity of extreme weather conditions. Increasing frequency of storm surges and rising sea levels exacerbate each other's effects. This together with the fact that many of the islands are rarely more than a couple of metres high is evidently potentially disastrous for several islands. These micro States are not amongst the most influential of States within the international arena. However, as detailed in chapter three, in order to try to achieve greater influence internationally primarily with regards to matters of climate change, small island States from around the world joined together to form

AOSIS. This alliance has enabled the islands to have a greater say in international relations than they would otherwise have had,[21] Nevertheless, even collectively, AOSIS represents a small power bloc. As was made clear in chapter six, the G77 group of States (of which AOSIS is a part) largely ignored the interests of AOSIS vis-à-vis renewable energies targets at the WSSD. The OPEC States who are also members of the G77 managed to overrule those countries that were in favour of targets for the increased use of renewable energy.

The vulnerabilities of small island States have been acknowledged by the international community. As mentioned in chapter six, SIDS are recognised in Agenda 21 as being 'ecologically fragile and vulnerable' (UNCED, 1992:66). Agenda 21 also states that

> The vulnerability and response options of small island developing States to global change and potential sea-level rise should be assessed. Also, based on precautionary and anticipatory approaches, response strategies should be designed and implemented to address the environmental, social and economic impacts of climate change and sea level rise and to prepare appropriate contingency plans. (UNCED, 1992:67)

More recently, as detailed in chapter six, the WSSD's Plan of Implementation called for SIDS to be given assistance to combat the adverse effects of climate change. This included the call for freshwater programmes; the possibility of the GEF being utilised in this regard was also mentioned.

There are many ways in which the actions of Germany may indirectly benefit small island States in general, and therefore, the Pacific Islands, but it appears there are not very many ways in which Germany provides them with direct assistance. Analysis of how Germany's climate change international relations may affect the island States will be conducted using the same order in which international relations have been examined in this study.

As explained in the introduction, Germany as an industrialised country has contributed to climate change. It is still emitting large amounts of GHGs and is, therefore, still contributing to climate change. Although it is impossible to apportion precise amounts or origins of GHGs to specific effects in particular areas of the world, it can be argued that Germany's industrial outputs are at least partially to blame for climate change and therefore, for the detrimental effects that are being experienced by the Pacific Island countries. It can therefore, be argued that Germany, along with other industrialised countries, has a responsibility to take action in order to mitigate climate change and to help other less economically fortunate countries to adapt to climate change. In fact in a section of Germany's third national communication to the UNFCCC that discusses adaptation measures, it is recognised that

> anthropogenic climate change is also an issue of equity between North and South. (Federal Government, 2002:148)

Germany is making efforts to limit its own GHG emissions and is also taking a leading role in trying to govern climate change. Chapter three detailed the evolution of the UNCED process, which included the establishment of the UNFCCC process. It was explained that AOSIS vociferously advocated the inclusion in the UNFCCC of the preventive principle; the precautionary principle and the polluter pays principle; principles that were already enshrined within German politics. This is an example of the fact that the positions of Germany and AOSIS often conflate. It is clearly the case that it is an advantage for AOSIS to have a country that is perceived to be one of the leaders in climate change politics arguing from a similar standpoint as itself. It can also be maintained that Germany's argument gains additional credence by having the support of States that are particularly vulnerable to climate change. Throughout this study, it has been seen that Germany has pursued progressive aims at global conferences. It can be argued that where Germany has succeeded in influencing the international agenda, Germany's actions have indirectly benefited countries that are particularly vulnerable to climate change.

The formation of the JREC is a prime example of Germany influencing the international agenda, even though its original aim of achieving agreed targets for the use of renewable energy at the WSSD had been defeated. As already discussed the pursuit of targets and timelines has been agreed by more than eighty States. If definite targets are introduced and adhered to, most States and all future generations will benefit as even if a low threshold agreement is reached it will result in a smaller amount of GHGs being emitted than would otherwise have been the case, which will in turn benefit the future prospects of the earth's environment. The term 'most States' has been used in the previous sentence as opposed to 'all States' because unless ambitious targets are agreed upon and met, it is likely that some of the most vulnerable Pacific Island States such as Tuvalu will not exist by the end of this century. It should perhaps be noted that of the eighty three (as of December 2003) coalition members, thirty four are countries that belong to AOSIS. Nevertheless, all twenty five EU Member States are coalition members, as are a number of other industrialised and industrialising countries, such as Brazil and Argentina. It can be argued that countries that constitute AOSIS are those that are likely to benefit most from any agreement and its following implementation and that they have the most to gain or to lose from the success or failure of this endeavour, thus they are the States that benefit most from Germany's initiative. It is however, also the case that the number of States having signed this coalition agreement is presented by various EU and German government sources as being proof of the potential for success of this initiative. The numbers are boosted quite dramatically by AOSIS countries having signed the agreement; they are therefore, giving credence to the initiative. It can also be argued that vulnerable States agreeing with the initiative is a benefit in that it adds some moral weight to the project.

Within this chapter it has been detailed that Germany funds the UNFCCC Bonn Fund, which finances the attendance of non-Annex 1 countries at UNFCCC meetings in Bonn (the countries that constitute AOSIS all fall into this category). This facility benefits all developing countries, including those most vulnerable to climate change. The fact that Germany also provides an expert to the Consultative Group of Experts on national communications from non Annex 1 parties can also be perceived as beneficial to vulnerable countries. If the German expert directly assists one of the Pacific Island States in the preparation of their national communication then this is clearly of benefit to them. However, even where the German expert is helping other developing countries, this is enabling the UNFCCC process to be implemented and therefore can indirectly benefit the small island States, this is especially so if the developing country concerned happens to be one that is rapidly industrialising.

During analysis of the GEF it was noted that Germany is an active member, it is a large financial contributor and it has a seat on the GEF Council, which approves projects. It has also been stated that whilst it can be surmised that Germany does influence the GEF to some degree, this influence is not measurable nor can it be determined whether or not Germany has had influence in the awarding of specific projects. Nevertheless, as Germany is particularly keen on working through the GEF it is pertinent to establish whether, and to what degree the Pacific Island countries benefit from GEF projects.

The GEF list of programs includes a project that directly addresses climate change in the Pacific Islands region. The project title is the Pacific Island Climate Change Assistance Programme (PICCAP), the islands covered are: the Cook Islands; the Federated States of Micronesia; Fiji; Kiribati; the Marshall Islands; Nauru, the Solomon Islands; Tuvalu; Vanuatu and Western Samoa. The GEF project fact sheet states that:

> PICCAP activities for this project are designed to:
>
> • Assist Pacific Island countries to identify climate change mitigation options.

- Enable Pacific Island countries to fulfil their UNFCCC reporting obligations regarding (a) development of inventories of greenhouse gas sources and sinks, (b) vulnerability to future climate change and sea level rise, (c) option for adaptation to climate change, (d) development of national implementation plans, and (e) communicating information. (www.gefweb.org/Outreach/outreach-PUblications/Project_factsheet/Asia_Pacific-paci-3-cc-undp-eng.pdf April 2004)

Foreseen benefits are to:

- Help region prepare for impacts of global warming
- Increase networking and exchange of information, expertise, and methodologies among countries
- Increase public awareness on climate change, sea level rise, and coastal management issues
- Share methods, outputs, and experiences at international level (Ibid)

PICCAP activities in practice have included a course run by the University of the South Pacific on climate change vulnerability and adaptation assessment.[22]

The GEF project list mentioned above lists several projects in numerous countries under the climate change section. In addition to the PICCAP, two other projects are mentioned that relate to Pacific Islands: Niue and Papua New Guinea, these are both detailed as climate change enabling activities. It should be said that there are also projects that cover small island States in the Caribbean and Indian Ocean. There are also a number of projects in China and other fast developing States. As has been stated previously these will indirectly benefit the Pacific Islands.

Another GEF funded project that directly assists Pacific Islands and contributes towards mitigating climate change is the Pacific Islands Renewable Energy Project (PIREP). Countries covered by PIREP are those included in PICCAP plus Niue, Palau, Papua New Guinea, Tonga and Tokelau.[23] In addition to reducing GHG emissions, this project will help islands sustainably develop. The Pacific Islands are currently heavily reliant on imported fossil fuels, which are especially expensive due to the transportation costs to these geographically remote islands, it is also estimated that less than thirty per cent of the islands populations have access to electricity.[24] The development of renewable energies in the Pacific Islands will therefore, have multiple effects, such as: widen access to electricity; save expensive transportation and import costs; and save GHG emissions, both from the burning of fossil fuels to produce energy in situ and also in transportation. PIREP aims to facilitate the promotion, development and commercialisation of renewable energy technologies in the Pacific Islands.

Whilst these GEF funded projects relate specifically to the Pacific Islands, the amount of German involvement cannot be ascertained. It should however, be remembered that Germany does contribute substantially to the GEF; it also places much emphasis on the work it conducts through the GEF.

It is worth noting that it is particularly pertinent for SIDS to implement adaptation measures; according to M.J. Mace,[25] it is very hard to get GEF funding for these purposes. Mace is also of the opinion that the required procedure for applying for GEF funds is problematic for SIDS. This is due to the fact that SIDS tend to lack the human resources required to complete the demanding application procedures. The point was made that in developed countries, such as the UK, there is a team of people working on climate change. Island countries may only have the resources to employ one person on climate change matters and they are often also responsible for other issues. Whilst it may be the case that a simplification of procedures would be of benefit to SIDS, these comments do not negate the fact that as detailed above GEF projects do appear to be of benefit to Pacific Island countries.

International relations where direct German involvement can be seen will now be addressed.

As mentioned in the multilateral section some German government funded workshops have been attended by countries within AOSIS. No documentation regarding these workshops has been discovered and it is therefore, impossible to ascertain their effectiveness. In early 2003, Karsten Sach[26] stated that the BMU had been approached regarding the possibility of funding a workshop for AOSIS, this was given a potentially favourable response but at that time no official proposal had been received. This indicates that Germany is willing to increase its assistance to AOSIS. However, as of May 2004 a proposal had still not been received,[27] the reason for this appears to be due to a change of personnel within the initiating body.

The dialogue partners of the partially Wuppertal Institute initiated project, 'South-North Dialogue – Equity in the Greenhouse' mentioned above, include AOSIS, represented by the Foundation for International Environmental Law and Development (FIELD). As previously mentioned although the project was initiated and is run in part by a German institute it is not a German government run project, although it is supported by them. It is too early to determine how influential this project may prove to be. However, it is demonstrative of discourse being conducted that aims to be equitable both in terms of its inclusiveness and in its goals.

Discussed previously was the utilisation of embassies to conduct outreach work in order to overcome the developing countries lack of trust with regard to the aims of Germany and the EU. This is a policy that if successful could result in future climate change agreements at the international level and hence would benefit all countries, including, and perhaps especially, the vulnerable island States.

In the bilateral relations section of this chapter it was explained that the BMZ has a list of cooperation countries with which it works. It was also explained that no SIDS appear on this list. The GTZ focuses its efforts in Africa, where Germany already has bilateral contacts and it is felt these can be utilised to maximal effect. When discussing this direction of focus, Philipp Knill[28] stated that projects within Africa can affect a few hundred million people, whereas projects with SIDS would only affect a few hundred thousand people. The point that projects conducted elsewhere in the world have indirect effects on the Pacific Islands applies yet again.

Although the small island States are not on the cooperation list, Knill asserted that Germany has helped fund an information system that connects the environment ministries of SIDS.

Another form of support that has not previously been mentioned is that the German government contributes to the funding of SIDSnet (small islands developing States network at www.sidsnet.org), an internet portal run by the United Nations Department of Economic and Social Affairs. Information on climate change and how this affects the SIDS can be found on this site as can numerous links, for example, to related documentation and island information. Again, any specific benefits that the islands receive from the German contribution to this site cannot be delineated.

It has been seen that many of Germany's efforts in the international relations of climate change may indirectly benefit the Pacific Island countries. It could be argued that were Germany not so proactive in this issue area that the situation with regards to international agreements and implementation could very possible be in a far more parlous state than is at present the case. This is of course conjecture, but if this is the case then the activities of Germany have substantially benefited the island States.

Direct German involvement with the island States is minimal. However, it can be argued that concentrating on assisting larger developing States, like China and Brazil, to develop in an environmentally friendly manner, has greater global benefits than assisting SIDS directly. Ultimately this will benefit the most vulnerable countries in the world, which includes the Pacific Islands.

When asked whether or not Germany's aims in, and the resulting outcomes of, climate change negotiations have a beneficial affect on SIDS, Bill Hare[29] of Greenpeace responded "absolutely". He expanded by stating that Germany is one of the main leaders in this area and that especially since COP6 Germany has taken the lead in trying to sort out funding for developing countries. Germany has taken the lead in trying to get financial commitments from the main European governments and Japan. He confirmed that internationally Germany is generally supportive of SIDS, and that a large part of the climate funds that will be available through the GEF as a result of the Marrakesh accords (agreement reached at COP7) is being given by Germany and that this will in part be directed to the small island States.

It is also the personal opinion of Jürgen Lefevere,[30] who has been FIELD's representative for AOSIS, that Germany's progressive stance in the international relations of climate change, ultimately has a positive effect on SIDS, including the Pacific Islands.

Overall, this section has argued that whilst Germany conducts minimal direct relations with the Pacific Island States and SIDS in general, that the international relations in which Germany engages indirectly benefits them. However, it should be noted that there appears to be gap between Germany's intentions and actual policies with regard to SIDS. In Germany's third national communication to the UNFCCC under the measures for adaptation section it is stated that

> Efforts in future must not be confined to reducing greenhouse-gas emissions and protecting and enlarging CO_2 sinks; more and more attention must be given to mitigating the negative impacts of climate change and to reducing relevant countries' vulnerability, via suitable measures for adaptation. This issue is of key importance especially for the most strongly affected countries – in Africa, central America and Asia and the group of small island States. (Federal Government, 2002:149)

The report goes on to say that the government is in discussion with partner countries with regard to such measures. As has already been stated SIDS are not designated as partner countries. It is clear from the above quotation and from the responses to relevant questions during interviews that Germany is concerned with the plight of SIDS. It was often the case in interviews that the respondents were convinced that something was being done in collaboration with SIDS, but that no specific evidence could be produced.

It is perhaps worth mentioning a 'Report on the Pacific Umbrella Initiatives post WSSD',[31] which details a variety of initiatives, a number of which can be related to climate change. Under the heading of 'Pacific Islands Adaptation', Australia and New Zealand are identified as potential partners. Germany is not mentioned under any section, although further dialogue with the EU is mentioned under the heading 'Pacific Islands Energy and Sustainable Development'; the EU energy initiative is noted as a related initiative.

An example of SIDS and Pacific Island States attitudes toward the type of international relations that Germany engages in can be found in facts such as them agreeing on the implementation of precautionary and polluter pays principles, and the JREC. It can also be found in responses to the ratification of the Kyoto Protocol. During the 2002 Pacific Island Forum annual summit a press statement was issued that included

> the Heads of State and Heads of Government …
> Warmly welcomed the acceptance of the [Kyoto] Protocol by Japan and approval of the Protocol by the European Community. (Pacific Islands Forum Secretariat, 2002)

It is also the case that Tuvalu has been planning legal action against the USA with regards to anthropogenicallly caused climate change. There is some reticence to include Australia in the action as they are traditionally substantial donors to the Pacific Islands, Tuvalu has however called upon Australia to do more to cut GHGs. Tuvalu's Finance Minister and former Prime Minister has stated that

> Tuvalu was not targeting nations like the European Union or Japan because they have accepted Kyoto. (Planet Ark, 30-08-02)

It has been argued in this section that Germany's efforts are largely beneficial to the Pacific Island countries, though such benefits are not quantifiable nor are they directly identifiable in the majority of cases. It is very difficult, if not impossible, to observe these benefits, just as it is impossible to directly attribute specific instances of climate change effects to a specific country's industrial output.

Conclusion

A commonality that is evident throughout the variety of international relations that have been examined in this chapter is their discursive and multi-participatory nature. Clearly, relations with and through multinational institutions such as the UNFCCC and the GEF require multi-participation and it can therefore, be argued, that all States that engage with these institutions engage in multi-participatory processes. This study and this chapter in particular, has demonstrated that Germany engages in these processes in a proactive manner that aims to facilitate the widening and deepening of these multi-participatory discursive opportunities, thus enabling more equitable relations, at least in terms of participation. It can of course, be argued that in some cases a small shift towards more equitable development may occur; for example the PIREP project will enable the Pacific Islands to benefit from increased access to electricity. German funded workshops for developing countries that are run by the multilateral institutions such as UNEP, as described above also fall into the category of facilitating more equitable participation in international relations.

The JREC and the International Conference for Renewable Energies have also been discussed. These were both German initiatives, and the latter was also hosted by Germany. It can be argued that the JREC was an initiative aimed at preventing complete defeat at the WSSD with regards to establishing targets and timeframes for increased renewable energy usage. It can also be argued that as Germany is a leader in renewable energy production the hosting of an International Conference for Renewable Energies could result in increased demand for German technologies and therefore, increased international trade. It could therefore, be argued that these initiatives were born of self-interest; and that Germany's actions are reflective of a traditional realist view of international politics. However, such an argument would be too simplistic. As has been shown throughout this study, Germany's aims, whilst obviously taking into account German interests, are also to find ways in which all can participate and benefit. When a more complex viewpoint is taken it can be seen that Germany's international relations fit more within a neo-liberal institutionalist framework, which allows for win-win situations. Neo-liberal institutionalism is also more akin to Germany's belief in the need for multi-participative discourse based international relations. The desire to maximise one's own interests whilst taking into account the interests of others is compatible with the argument put forward by Habermas, that whilst humans can act strategically, their actions are also influenced by inner values and social norms.

The inclusive South-North dialogue discussed above is also an example of the German aim to include multiple parties in equitable dialogue aimed at reaching an equitable framework for future negotiations. The fact that this dialogue is not run by the German government, but that it is supported by it, can be seen as indicative of discourse ethics being put into practice, where interested and affected parties have a right to have their say and to be listened to.

Mentioned in both the multilateral and bilateral sections was the decision to use embassies to systematically conduct outreach work. Although these are bilateral efforts they are multi-participatory in that embassy staff, host government personnel, German companies and local media and hence local citizens are included in such discourse. It is also the case that such bilateral relations are being conducted with numerous countries, indeed Germany is trying to persuade other EU Member States to undertake similar activities, and hence it could be said these bilateral relations are being conducted on a multilateral basis. Whilst the use of embassies in this way is aimed at persuading the host country of the efficacy of Germany's aims, this can still be seen as part of a discursive process. This is because Germany is also willing to listen to the concerns and suggestions of the country concerned, with the aim of coming to some agreement prior to large international negotiations. It also demonstrates a willingness to communicate and gain agreement through consensus.

Development cooperation as the name suggests is development assistance that is conducted in cooperation with the country concerned. Clearly for both parties to agree on the benefits of projects and hence on their implementation entails discourse. However, some issues of German economic self-interest apply to the transfer of technologies to industrialising countries, such as China. Indeed some aspects of power politics could be said to be relevant, for example in relation to cooperation development in Africa, where it has been said that Germany has some leverage to encourage the implementation of adaptation measures. Nevertheless, development cooperation projects are not imposed upon countries but are managed on a cooperative basis. These projects will assist developing countries to sustainably develop and will include climate change considerations that Germany deems necessary. Hence, the developing country benefits and ultimately so will countries around the world. The analysis regarding the combination of self-interest, participative discourse and the interests of others being compatible with the works of Habermas, put forward in the paragraph regarding the International Conference for Renewable Energies, also applies here.

The latter section of this chapter examined the effects that Germany's international relations have on Pacific Island States as they are particularly vulnerable to climate change. It was seen that minimal direct assistance was given to the Pacific Island States by Germany. However, it was concluded that the proactive stance that Germany has been shown to take in international negotiations, international institutional processes such as the UNFCCC and the GEF, and the various other multilateral and bilateral relations that have been discussed, do benefit the Pacific Islands. These benefits result from any climate change mitigation brought about by reduced GHGs whether this is due to the implementation of international commitments or from the promotion and transfer of technology with regard to the use of renewable energy and increased energy efficiency. It is also the case that Germany's commitment to inclusive discourse facilitates the participation of these States in the ongoing climate change debate.

These findings show that Germany influences the climate change international relations process and that Germany's propensity to engage in inclusive discourse translates into actions that often facilitate multi-participation, i.e. that enhance the implementation of participatory justice.

Each chapter in this study has shown that Germany takes seriously its commitment to manage the effects of climate change; this chapter has consolidated this finding. The task of the next chapter will be to conclude this study and to carry out final analysis of the study as a whole.

8

CONCLUSION

This study has examined Germany's role and influence in the international relations of climate change. In order to accomplish this, Germany's domestic policies and politics, and the impacts that these have on international relations have been assessed. Exogenous events that have affected Germany's domestic politics have also been noted. The way in which Germany conducts climate change international relations has been analysed and the influence it manages to achieve has been assessed. Both Germany's political system and international relations tend to be multi-participatory in nature, with agreements being sought that are arrived at by consensus. It is argued that agreements are more likely to be successfully implemented if they have been arrived at in a manner that is perceived as being just, i.e. that the viewpoints and considerations of all concerned have been taken into account. Analysis regarding the implementation of participatory justice has utilised Jürgen Habermas' work on discourse ethics.

Climate change is a phenomenon in which issues of justice are embedded. Distributive and inter-societal injustices exist in terms of the benefits and detriments of industrialisation. Benefits include economic and political power; detriments include the adverse effects of climate change, the severest consequences of which are felt by those countries that have not benefited from industrialisation and are therefore, least able to afford adaptation measures and cannot contribute greatly to mitigation efforts. The industrial causes of climate change act slowly, thus intergenerational justice considerations also apply. Although justice issues permeate climate change politics, it is often the case that this is implicitly so. Whilst acknowledging a variety of global justice issues, this study has focused on participatory justice in the form of discursive inclusiveness. As explained in chapter two, Habermas' holds that decisions should be made through inclusive discourse and communicative rationality. Discourse should ideally include all actors that have the communicative ability and desire to participate, everyone should have equal opportunity to put forward their point of view and the result should be rationally decided upon, in other words, the best argument should prevail. In addition to the pertinence of considerations of participatory justice due to the consensual nature of German politics and the international relations of climate change, it is argued that it is a necessary foundation for successful climate change governance that it is perceived as being just and hence acceptable and workable. Without such acceptance intergenerational and inter-societal justice cannot be fulfilled.

Chapter three began the empirical body of this study; it provided an historical backdrop to the more contemporary investigation that has formed the main focus of this study. This scene-setting chapter examined the evolution of climate change politics and policies in Germany and internationally through the UNCED process. It was seen that Germany introduced its first Environment Programme in 1971, the principles contained therein have been considered in all subsequent environmental policy-making. These principles: precautionary, polluter-pays, and industry-government cooperation are examples of justice considerations being intrinsic. As explained in chapter three, the precautionary principle can be related to intergenerational justice; the polluter-pays principle relates to retributive justice, i.e. taking responsibility for distributive injustices; and industry-government cooperation can be seen as part of the process leading to participatory justice. It was argued that the institutionalisation of these principles can be related to Habermas' argument that generalisable interests can become universal norms.

It was shown that at various times German environmental policy-making has been in response to, or reinvigorated by, popular opinion and demand. This was the case during the 1980s following the popular realisation, and subsequent outcry, that air pollution was causing *Waldsterben.* It was also the case that popular concern regarding the safety of nuclear power, which was exacerbated following the Chernobyl nuclear accident, led to increased support and electoral success for the Green Party. This in turn led to 'green' issues being taken on board by the main parties. The SPD adopted many of the Green Party's environmental polices including that against nuclear powered energy. It was therefore, concluded that Germany's political system and its propensity for discursive inclusiveness has allowed popular ideas and values to be considered and thus become a part of the decision-making process. This process is akin to the call for decision-making through multi-participative discourse and communicative rationality that is called for in Habermas' discourse ethics.

Progressive contemporary climate change related policies include: the ecological tax reform; the renewable energy act, the 100,000 (solar) roofs programme; and feed-in rules that ensure that renewable energy is purchased by the grid at a premium price. Germany is a leader in the production of wind energy and in the development of solar power stations. It has been argued that:

- The inclusiveness of Germany's federal system has allowed 'green' issues to become mainstream concerns and that this has led to the development of effective domestic environmental policy-making.
- This has given credibility to Germany's aims in the international arena, hence enabling Germany to become a leader in the international relations of climate change. Successful domestic policies can be seen to contribute to discourse by showing that a good and potentially best argument is being put forward by Germany.
- There is domestic–global reflexivity; i.e. what happens at each level impacts on the other. Examples include the transnational issues of air pollution causing *Waldsterben* and the Chernobyl nuclear accident, both of which caused public concern and political action, resulting in effective domestic environmental policy-making which in turn has led to Germany being a leader in climate change international relations. Another example can be found in increased energy efficiency in response to the global oil crisis, this led to the acceptability of policies to further increase efficiency of energy use, in turn leading to international credibility and leadership. The Chernobyl accident and the oil crisis can be seen as unintentional exogenous contributions to discourse, or at least events that have engendered discourse and that have led to an ongoing learning process.

Although chapter three did not particularly address Germany's role in the international relations of climate change, it was seen that Germany has exerted influence in the international arena. According to Von Weizsäcker, Klaus Töpfer who was Environment Minister from 1987 to 1994 played a leading role in environmental policy-making within Europe, and in the success of UNCED. It was also seen that Germany was amongst the more progressive States at UNCED and that the environmental principles that Germany established in 1971 are reflected in the 1992 Rio principles. This does not establish a direct causal link, but it can be assumed that Germany was amongst a group of States that advocated these principles.

As indicated above, it has been argued that discursive inclusiveness, or in other words, participatory justice has resulted in the ideas and values of the populace being taken on board by the government, and acted upon. It has been shown that systems are in place within Germany that enable the opinions of numerous sectors of the population to be made known, and therefore considered in environmental policy-making. The consensual nature of the federal system within Germany allows for a wide variety of ideas and opinions to be considered. It is also the case that structures have been put in place that enables professional research to be conducted and for the government to be informed of the results, examples of which include Enquette Commissions (these were explained in chapter three), and the WBGU (mentioned in chapter three and discussed more fully in chapter four). In chapter four it was also seen that in the formulation of the national sustainability strategy the general population had the opportunity to contribute their opinions through an on-line facility established by the Council for Sustainable Development. It can thus be seen that through discursive inclusiveness the ideas and values of many can be considered, discussed and consensus reached. In this way, ideas that through rational discussion have been deemed to form the 'better argument' are reflected in the official German position with regards to both the domestic and the international arena. It is argued that this is reflective of Habermas' discourse ethics in action; it is not a perfect theoretical reflection, but nevertheless, participatory justice is, at least in part, being implemented.

Chapter four built on the information contained in chapter three and examined the processes by which Germany arrived at its aims for the WSSD and hence COP8 which followed shortly thereafter. It was argued that Germany's aims for the WSSD were a combination of the evolutionary processes and policies that were analysed in chapter three and of a multiplicity of inputs from a variety of interested parties in the few years leading up to the conference. The positions of selected research institutions, and non-governmental organisations that were deemed to be relevant in that they were taking an active role in producing information that was shown to be considered in the governmental decision-making process were analysed. It was seen that the policies advocated by the various organisations often overlapped. This is consistent with Habermasian thought, as through ongoing discourse, consensus is sought and over time more similarity of positions is therefore, likely. The input of the business/industry sector was also assessed as was information from a variety of government sources. It was argued that whilst not all sectors were, nor could be, completely satisfied with the government position, that the official stance was reflective of the various inputs and that this was demonstrative of discourse ethics in action.

Analysis in chapter four argued that

- Germany's successful domestic policies, for example, with regard to the promotion and development of renewable energy, feed through to Germany's aims in international relations. As mentioned above, policies that are proven to be effective impart credibility to Germany's contribution to international discourse and in this manner influence can be achieved.

- Germany's aims for the WSSD were developed through an ongoing and discursively inclusive process. Participatory justice has been shown to be implemented to a large degree in the formulation of Germany's aims.

Chapter five examined Germany's relationship with the EU in terms of climate change related politics and policy evolution. It has been shown that Germany has actively participated in the environmental policy development of the EU. It was argued that Germany was influential in the adoption of the Large Combustion Plant Directive and that it was active in promoting the adoption of the precautionary principle at the European level. The nature of the EU means that inclusive discursive processes are a necessity in policy formation.

At conferences such as the WSSD and COP8, the EU presents a united position. Thus chapter five also assessed Germany's role within the EU's preparatory process. The meetings at which Member States discuss the EU stance are not minuted and thus a lack of transparency has prevented definitive causal links between Germany's input and that of the EU being proven. Comparison of Germany's stance and that of the EU, together with assessing information gained during interviews with a variety of people, has enabled the conclusion that Germany influences the position taken by the EU. Comparison of German and EU aims shows that whilst they are not identical, there is a great deal of confluence. A difference can be seen in that the EU advocated the strengthening of UN bodies and increased cooperation between these and the WTO, whilst Germany would like to see the creation of a strong WEO that would be a counterweight to the WTO. Information gained from interviews supports the contention that Germany's influence is the result of a variety of factors and includes: the personal agency of Karsten Sach of the BMU, who is an active and respected participant in the EU process; structural factors such as the credibility imparted to Germany through its successful and progressive domestic policies; and perhaps from the fact that Germany is a major financial contributor to the EU.

It was argued that the process of agreeing on a united EU position, which entails regular meetings between negotiators from Member States and the Commission, is itself reflective of the requirements of discourse ethics, and that Germany actively participated in this process.

Chapter five also put forward the assertion that energy was included on the agenda for the WSSD in large part due to the EU, France, Austria and Germany. Contributing to energy's inclusion on the WSSD agenda was of major importance as the promotion of renewable energy has been a major strategy of Germany's and one where it has since had influence in the international arena.

Analysis in chapter five has argued that

- Germany influences the climate change international relations process; this has been shown at the EU level and in agenda setting for the WSSD. This influence is achieved through discursive processes.
- Germany's domestic policies give credibility to its international relations aims and therefore, contribute to the discursive process and to Germany's influence.
- Germany's aims are mediated by the international process; the process of agreeing on a unified EU stance necessitates this. This is a feature of discursively agreed consensus or compromise.
- There is reflexivity between the German and EU levels; EU agreed policies give extra weight to the BMU's argument at the domestic level. Germany participating in EU level discourse and policy-making has the effect of the EU, in many ways, being included in domestic policy-making discourse.
- Discursive inclusiveness is a major factor in EU policy formation.

Events at the WSSD and COP8 were analysed in chapter six. Large German delegations were present at both conferences, as they are at all international conferences. This enables Germany to keep track of unfolding events and to be involved in discussions with a variety of people, thus providing the opportunity for influence.

In addition to Germany's active involvement at these conferences and in particular at the WSSD, Germany provided substantial financial support for the staging of the WSSD. The German government also funded the attendance at the conference of various organisations including various African NGOs. This is another example of Germany contributing towards participatory justice, or in Habermasian terms, enabling multi-participative and inclusive discourse to be engaged in. Germany also had a stand at the cultural and side events village where many events were held, one of which promoted the role of solar power in sustainable development. This can be seen as widening access to Germany's contribution to discourse. Whilst an element of self-interested promotion may well have been involved in this and similar events, it is also the case that the promotion of renewable energies, if taken up, will benefit the climate and therefore future generations.

Germany and the EU wanted time-bound targets to be agreed at the WSSD for increased renewable energy use. This did not happen, due mainly to opposition from the USA and the OPEC States. However, energy was discussed at the conference, in part due to the efforts of Germany, and the Plan of implementation does recognise that the sustainable development of energy supplies is a necessity and the increased use of renewable energy is called for. Whilst Germany's aims were not met in full, it is evident that Germany's discursive efforts did influence the international relations process. Germany continued to influence events following the failure to reach agreement on energy targets. The Johannesburg Coalition on Renewable Energy was formed following a German initiative that led to the EU submitting a declaration of like-minded countries for promotion of renewable energy, which includes a commitment to work toward the introduction and implementation of renewable energy targets. Chancellor Schröder also announced that Germany would host an International Conference on Renewable Energies. Germany's active role in promoting increasing renewable energy use is an example of their domestic policies giving them credibility in the push for international agreements; they are after all 'practicing what they preach'. Zeal for environmental action to be taken on a global scale can be argued to be due to the belief that such action is a necessity to ensure a fair and viable future for all; the reflection of the values and ideas that are important to individuals being sought for all. It can also be argued from a more cynical viewpoint, that in many ways the adoption of domestic policies that impose strict environmental conditions necessitate a foreign policy that enthusiastically promotes similar policies to be adopted at an international level so as to prevent domestic isolation and competitive disadvantage. If progressive policies are adopted at an international level, Germany is then at a competitive advantage being in a position of technological advancement and thus able to sell its expertise abroad. The likelihood is that both factors have a bearing on Germany's aims. This is consistent with the works of Habermas, in which it is recognised that actors consider strategic concerns but that actions also take into account values and social norms.

COP8 provided no definitive evidence of Germany's influence. However, it was argued that the inclusion in the Delhi Declaration of the call by those countries that have ratified the Kyoto Protocol for those that have not ratified to do so, was mainly due to the influence of the EU, France and Germany. It has been argued that this was achieved in the face of obstructive behaviour on the part of the USA and the OPEC States. It was also pointed out that for this to have been achieved indicates that elements of ethical discourse were evident. Whilst the participatory ethics of the USA and OPEC were called into question, it is the case that the conference framework allowed for a compromise text calling for ratification of the Kyoto Protocol.

Other issues where Germany's aims coincided with the outcomes of the WSSD and COP8 and where influence may have occurred, but where no causal link has been proven include the reaffirmation of the Rio principles and the agreed need for: increased renewable energy use; increased efficiency of energy use; financial and technical assistance for developing countries; increased capacity building in developing countries, with the assistance of international cooperation; public/private partnerships; the implementation of transport policies that minimise GHG emissions; national promotion of sustainable development; the strengthening of UN bodies and for increased cooperation between them, i.e. between the GEF and the WTO; and environmental assessments.

In chapter six it was demonstrated that

- Germany influences international relations that relate to climate change and that this is achieved by engaging in multi-participative discursive processes.
- Germany's domestic policies regarding renewable energy gave credibility to its aims at the WSSD, and Germany's ratification of the Kyoto Protocol imparted influence at COP8. As previously argued, Germany's successful domestic policies contribute to the perceived validity of Germany's contribution to discourse.
- Germany's aims are mediated by the international relations process; the evolution of events pertaining to energy substantiates this claim. As already mentioned, this can be expected of multi-participatory discursive processes.
- There is domestic–global reflexivity; the energy issue also corroborates this contention. The domestic effecting the global has been previously detailed. Decisions regarding energy at the global level impact on how Germany's aims are pursued. The delayed ratification of the Kyoto Protocol also impacted on Germany's activities, for example, in terms of continued encouragement for this aim and in the content of continuing negotiations (i.e. COP9 in 2003). Reflexivity is a characteristic of discursive inclusiveness.
- It was argued that the discursive nature of the WSSD and COP8 meant that the use of discourse ethics to aid analysis was appropriate. It was also noted that there had been criticism of the lack of progressiveness of both conferences, and that this coincided with the characteristics of agreements reached by consensus through an openly discursive process.

The WSSD and COP8 were major international conferences. Analysis of the WSSD and COP8, together with Germany's policies, aims and actions in relation to these conferences, is important in determining Germany's role in the international relations of climate change. However, to enable greater understanding of Germany's influence, international relations that continue on a regular basis away from the focus of such major conferences need to be examined. Ways in which Germany conducts climate change international relations on a more routine basis were analysed in chapter seven.

The location of the UNFCCC secretariat in Bonn and the possibility of this providing any additional influence to Germany was examined, as was work that Germany conducted with, and through the UNFCCC process (outside of the conferences). Subtle influences probably occur due to the positioning of the secretariat, for example the UNFCCC staff live in Germany and therefore, become aware of important domestic issues. Everyday discourse may have imperceptible effects. Germany sources the Bonn Fund which enables various workshops to be undertaken and for the attendance at such events by developing country representatives that would otherwise not be able to participate. Germany also provides one of the group of experts that assist developing countries in completing their national communications under the UNFCCC process. Germany does, therefore, exert subtle influence through the UNFCCC process, but in a way that widens participation in the process. Thus it can be argued that Germany helps toward the implementation of participatory justice.

Germany is keen to work through the GEF, and this relationship was also examined. Germany actively participates in the GEF and is a major financial contributor to it. Apart from the contention that a GEF unit was streamlined following a suggestion from Canada and Germany, no evidence of influence has been found. Nevertheless, it was argued that Germany probably influences the GEF process, but again this is in a very subtle and unobservable manner.

Other multilateral relations were examined, for example: the funding of workshops; the hosting of the International Conference for Renewable Energies; and the BMZ and GTZ supported 'South-North Dialogue – Equity in the Greenhouse' project. Each of these involves multi-participative discursive processes.

Although multilateral relations constitute the major parts of Germany's international relations, bilateral relations do occur and these were also investigated.

One of Germany's strategies to try and 'get on board' developing countries prior to future conferences is to utilise German embassies around the world by engaging the host country in discussions. Such discourse will enable Germany to understand developing country concerns and also help the host country to understand Germany's aims. This discursive process is underway in Thailand and it seems to be proving beneficial to both parties.

Bilateral relations between Germany and: other EU Member States; Japan; the USA; Brazil; China; and India were discussed, as were relations with developing countries through GTZ projects. The form of these relations varies, but all include discourse. This can even be argued to be the case in instances where other considerations also apply. Economic self-interest issues play a part in trade and the transfer of technologies to other countries, and elements of power politics could apply in development cooperation projects. Nevertheless, it is not the case that these other considerations rule completely; discourse and consensus is a requirement in what trade and technological transfers take place and in the acceptability and implementation of development cooperation projects.

Chapter seven also included analysis of how Germany's efforts affect Pacific Island countries that are particularly vulnerable to climate change. It should be pointed out that Germany, as an industrialised country bears part of the responsibility for human induced climate change and therefore, part of the responsibility for the damaging effects being experienced by the Pacific Islands. However, it can be argued that Germany's efforts in climate change international relations indicate that it is taking on that responsibility and is making enormous efforts to ensure that future development changes course to a more environmentally sound pathway. Minimal direct relations exist between Germany and the Pacific Islands, and so it is difficult to outline any specific benefits that the Pacific Islands receive from Germany's actions. However, the proactive and progressive stance that Germany takes in international negotiations, together with the numerous and varied international activities that Germany engages in, lead to the conclusion that Germany's climate change related international relations indirectly and ultimately benefit the Pacific Islands.

Argument in chapter seven demonstrated that

- Germany influences the climate change international relations process and it does so in an increasingly inclusive discursive manner. The expansion of participatory justice can lead to a slightly more equitable spread of the benefits of development.
- Germany's domestic policies such as those on renewable energy give credibility to its international relations discourse and aims.
- Germany's aims are mediated by international discursive processes; for example working through the UNFCCC and GEF.

Questions were identified in the introduction that could be asked to help Habermasian analysis of the German political system and the German approach to international climate change relations. Whilst these questions have not been specifically referred to throughout the text, they have been addressed. It has been shown that

- Multi-participation in policy-making is encouraged and implemented.
- Discussions are inclusive of numerous interested and affected parties. To be inclusive of all parties is an ideal, which it can be argued is virtually impossible to fulfil.
- As decisions are largely consensual it would appear that decisions are arrived at through rational-decision making. If this were not the case, consensus or compromise would be much less evident.

- When principles are agreed to be justifiable and universalisable, they are institutionalised. This can be seen by the enshrining in law of the precautionary, polluter-pays and industry-government cooperation principles.

Leading on from these questions it was also stated in the introduction that from a Habermasian point of view one might expect

- German climate change related policies to be the result of rational decision-making arrived at through discourse that includes a multiplicity of inputs and considerations.
- Germany's international relations of climate change to encourage and facilitate participative and just discursive processes.

It has been shown that Germany's climate change politics and international relations do possess these attributes.

Difficulties that have been encountered in this study have involved establishing causal links between discourse, documentation and outcomes. Information gleaned from some interviews and follow-up correspondence established that the documentation examined in the formation of Germany's stance for the WSSD was considered. Similarities between documentation and governmental positions and speeches were also pointed out. As there were no minutes taken in meetings in which the formation of the EU stance was discussed, it is impossible to determine a definitive causal link between Germany's position and that of the EU. Nevertheless, analysis of the two positions, together with information gained from interviews established that Germany was active and influential in this process. Comparison of German and EU aims for the WSSD and COP8 with the outcomes of these conferences, together with analysis of reports from the conferences and information from interviews, enabled assessment of influence or in many instances probable influence. These difficulties may be perceived as a weakness of this study; however, it is in the nature of decision-making through multi-participative discourse that it is extremely difficult to apportion specific amounts of influence to certain participants. This is especially so because continuing discourse aimed at reaching consensus is bound to lead to a similarity of positions. It can be argued that these difficulties actually support the premise of using Habermas' discourse ethics in analysis.

Analysis based on Habermas' discourse ethics can therefore, be considered a particular strength of this study. It has facilitated analysis of, and therefore, understanding of how multiple interests are considered in decision-making processes, leading to consensually agreed upon interests and aims. Thus Habermasian analysis has led to a greater understanding of how interests have been formed and legitimised. This study has shown how ideas and values within Germany have fed through to aims and actions taken at the domestic, EU and international levels.

A further strength of using Habermas' discourse ethics in analysis is that it has shown that participatory justice provides a framework through which decisions can be reached that are perceived as just and are therefore more likely to be implementable. This could be used to positive future effect.

A major theme of this study has been the German propensity to conduct politics and international relations through multi-participatory discourse. International relations, particularly those that occur within the confines of global conferences such as the WSSD and COP8, have a discursive imperative. However, not all countries engage in negotiations in a manner that coincides with discourse ethics as laid down by Habermas. It has been argued that the conduct of the USA and OPEC at the WSSD and particularly at COP8 was obstructive, with discourse being used in a deliberately disingenuous manner. However, it appears to be the case that Germany genuinely uses discourse as a means of trying to reach agreement through the force of the better 'truthful' argument. It has also been seen that Germany is active in widening the inclusiveness of discourse. A belief in consensual agreements reached through discourse does not mean that Germany does not aim to influence international relations. Discourse can be used to persuade others that a particular course of action is optimal. The active use of discourse conducted in an ethical manner points to a belief that overall their aims and arguments are right and just. It also shows a willingness to listen to others and consider their viewpoints and concerns. It is pertinent to reiterate the point made in chapter's five and seven, that Germany's propensity to conduct international relations in a discursive manner, rather than through traditionally realist power politics, was seen by a number of interviewees as being the result of Germany's history. This observation indicates the possible origins of German political culture and it in no way negates, and it may in fact help to explain, Germany's belief in discourse, and participatory justice.

To sum up, investigation of Germany's international relations of climate change has shown that Germany does exert influence in this issue area, and it does so mainly in a positive and just manner.

This research could be further developed by building on Habermas' discourse ethics and the nature of community. Realist theory only recognises the notion of community at the national level. It can be argued that such an assumption is inadequate. It can be seen from this study that there are in fact multiple overlapping communities, for example: the German national community, the EU, and within and crossing these, the EU environmental working group. At the truly international level one can discern amongst others: national, environmental, and international institutional (i.e. UNFCCC, GEF, UN) communities. Whilst separate communities can be recognised it is also clear that these interact and overlap. Discourse is the key to the forming and strengthening of these communities and it can be argued that ethical discourse is paramount in the successful achievement of the UNFCCC objectives.

Future research could examine the development of communities of interest. It has been shown that many of Germany's objectives coincide with those of AOSIS. The development of cooperation between these actors could be researched in order to determine how to maximise the utility of joint aims and actions. The forming, or strengthening, of such communities of interest would require multi-participative discursive processes of the type advocated by Habermas.

NOTES

1. Introduction

1 Emissions that contribute to global warming.

2 For further information on neo-liberal institutionalism see Kegley, Charles, W. Jr & Wittkopf, Eugene, R. *World Politics: Trend and Transformation. Seventh Edition.* (Boston & New York, 1999).

3 For further information on realist theory see Kegley, Jr & Wittkopf. *Trend and Transformation.*

4 See Habermas, Jürgen. 'The Tasks of Critical Social Theory' in C. Calhoun, J. Gerteis, J. Moody, S. Pfaff & I. Virk (eds) *Contemporary Sociological Theory.* (London, 2002) pp.377-400, & How, Alan. *Critical Theory.* (Basingstoke & New York, 2003) p.155.

5 See Grix, Jonathan. *Foundations of Research.* (London, 2004) p.78.

2. Theory Overview

1 Similar wording is used in Collins dictionaries.

2 See Beuermann, Christianne & Jäger, Jill. 'Climate Change Politics in Germany: How long will any double dividend last?' in T. O'Riordan & J. Jäger (eds) *Politics of Climate Change: A European Perspective.* (London,1996) & Sturm, Roland. 'Continuity and Change in the Policy-Making Process' in G. Smith, W. Paterson & S. Padgett (eds) *Developments in German Politics 2.* (Basingstoke, 1996). Following chapters will expand on consensual processes in Germany.

3 The full text of the Declaration can be found on the UN website. www.un.org/Overview/rights.html June 2002.

4 All affected parties having the opportunity to contribute to the discussion, all contributions being listened to, and decisions being made through rational choice, i.e. that the force of the better argument prevails.

5 Agenda-setting is identified as one of three faces of power by Lukes, Steven. *Power: A Radical View* (London, 1974).

3. Evolution of Climate Change Politics and Policies

1 The public became aware of massive tropical deforestation and the fact that forests act as 'carbon sinks', that is they take in carbon dioxide and give off oxygen. Carbon dioxide (CO_2) is a major contributor towards global warming and climate change.

[2] For further information see Mayer, Margit and Ely, John. (eds) *The German Greens: Paradox between movement and party*. (Philadelphia, 1998).

[3] It is stated that 'Data refer to 2000 or to the latest available year from 1996 on.' OECD (2002:7).

[4] Töpfer went on to be Chairman of the UN Commission on Sustainable Development between May 1994 to May 1995, at the same time he was Federal Minister for Regional Planning, Building and Urban Development and Coordinator for the Transfer of the Parliament and Federal Government to Berlin and Compensation for the Bonn region. From February 1998 until June 2006 he was UN Under-Secretary General and Executive Director of United Nations Environment Programme and Director-General of the UN office at Nairobi.

[5] Following the installation in November 2005 of a Grand Coalition between the CDU/CSU and the SPD led by Chancellor Angela Merkel of the CDU, Wieczorek-Zeul retained her post. SPD representatives became Foreign and Environment Ministers.

[6] CO_2 - Carbon dioxide, CH_4 - Methane, N_2O - Nitrogen Oxide, H-CFC - Hydro-chlorofluorocarbon, CFC - Chlorofluorocarbon, SF_6 - Suplhur hexaflouride.

[7] Willetts, Peter. 'From Stockholm to Rio and beyond: the impact of the environmental movement on the United Nations consultative arrangements of NGOs', *Review of International Studies* 22 (1996) pp. 57-80. contests this claim, arguing that in fact the UN Conference on the Human Environment in Stockholm in June 1972 holds this honour. The Stockholm conference may have been the turning point toward global conferences, attending NGOs numbered some two hundred and fifty, however this is less than a quarter of the numbers attending UNCED.

[8] For further information see Willetts, 'From Stockholm to Rio and beyond' pp. 57-80 and www.ciesin.org November 2001.

[9] Predominantly a scientific conference, The First World Climate Conference was held in 1979 and at which it was recognised that climate change was a serious problem. Scientists from a range of disciplines attended the conference. The Second World Climate Conference was attended by representatives from one hundred and thirty seven States and the European Community.

[10] AOSIS now has a membership of 43 (4 of which are observers) and is described as "a coalition of small island and low-lying coastal countries that share similar development challenges and concerns about the environment, especially their vulnerability to the adverse effects of global climate change." www.sidsnet.org/aosis/main-aosis.htm November 2001. Germany contributes to the funding of Sidsnet.

[11] Interview 03-03-03.

[12] These detailed reports can be found at unfccc.int.

4. Pre-World Summit on Sustainable Development

1 Personal correspondence with Jessica Suplie of the BMU and Philipp Knill of the German Embassy La Paz, ex of the Ministry for Economic Cooperation and Development (BMZ), November 2004. Although the BMU is the lead ministry with regard to climate change, the BMU and BMZ are jointly responsible for the Rio process of which the WSSD was a part.

2 Interview, 21-02-03.

3 Personal correspondence with Philipp Knill, November 2004.

4 For details see Heinrich Böll Foundation, *From Rio to Johannesburg: Contributions to the Globalization of Sustainability: World Summit 2002 Johannesburg Papers No. 5.* (Berlin, 2001) p. 25.

5 2012 is when the Kyoto Protocol's first commitment period ends.

6 Personal correspondence with Judith Vorrath of the SEF, November 2004.

7 Personal correspondence with Philip Knill, November 2004.

8 Personal correspondence with Jessica Suplie, November 2004.

9 Personal correspondence with Jessica Suplie, November 2004.

10 In an interview (11-03-03) Christoph Bals from Germanwatch expressed the view that the independence of the Forum was not at all compromised by being partially funded by the BMU and the BMZ, as these government departments understand that their positions within the government are strengthened by NGO criticism.

11 Personal correspondence with Jessica Suplie and Philipp Knill, November 2004.

12 For further information see www.unece.org/env/pp/welcome.html November 2002.

13 For further information see www.rio-10.de/berliner_aktionstage.html August 2002.

14 Interview 11-03-03.

15 For further information see www.rio-10.de/presse/pm_offener_brief060802.html August 2002.

16 For further information see www.rio-10.de/rioprozess/texte/kanzleramt_antwort_%20130802.pdf August 2002.

17 Interview 22-04-03.

18 In the third replenishment of resources Germany provided 11% of GEF contributions, behind the USA's 20.86% and Japan's 17.63%. The only other countries that provided more than 5% of funds were the UK's 7.93% and France's 6.81%. For further information see Global Environment Facility, *Summary of negotiations on the third replenishment of the GEF Trust Fund.* (2002) p. 44. The report also notes that the Global Environment Facility was established as a result of a Franco-German initiative.

5. Germany and the European Union

1 The EU has been granted observer status within the UN, however, it was given participation status for UNCED, but it has to re-negotiate this position for each subsequent conference i.e. for the WSSD. However, the EC as a regional

economic integration organisation has been recognised by the UNFCCC; it is a party to the convention and has continuing rights to participate at UNFCCC COPs. The EC does not have a separate vote from its Member States. (Vogler, John. 'The European Union as an Actor in International Environmental Politics', *Environmental Politics.* 10:2 (1999) pp. 33-34); unfccc.int/essential_background/convention/convention_bodies/negotiating_groups/items/2834.php November 2004.

2 Interview 21-02-03.

3 Interviews Karsten Sach 08-04-03 & Simon Crabbe 12-09-03.

4 Personal correspondence with Sebastian Oberthür during November 2004. Sebastian Oberthür has been Germany's representative in the EU legal expert group which is set up under the Working Party International Environment (Climate Change).

5 Interview 20-02-03. As well as being a member of an EU expert group (as detailed in the previous footnote) Sebastian Oberthür is a Senior Associate with Ecologic in Berlin, which is an institute for international and European environmental policy.

6 Interview 25-03-03. Bill Hare is International Climate Policy Director of Greenpeace.

7 Denmark also has to reduce emissions by 21%, Luxembourg is the only country that has to reduce by a higher percentage, which is 28%. Portugal has the greatest allowance to increase emissions, though Greece is not far behind with 25%. More information can be found at http://europa.eu.int/comm/environment/climat/gge_press.htm July 2003.

8 Interview with a member of the Directorate General of the Environment of the European Commission, 24-02-04.

9 Interview 12-09-03.

10 Interview 12-02-03.

11 Interview 17-04-03.

12 Interview 08-04-03.

6. WSSD and COP8

1 In an address to the round table on climate change and sustainable development at COP8 on 31-10-02, Trittin talked of common but differentiated responsibilities. More detail is given later in this chapter.

2 The CSD9 report can be found at www.un.org/esa/sustdev/csd/ecn172001.19e.htm February 2004.

3 According to a European Environment Agency Briefing in February 2004, Germany's subsidies to the coal industry exceeds 4 billion euros, it is also stated that in 2001 Germany's support for renewable energy exceeded 1 billion euros.

4 Telephone interview 27-05-03.

5 The Presidency of the EU was at the time held by Denmark..

6 Schröder came from behind in the opinion polls to win the election. This is largely attributed to the stance he took against going to war with Iraq; however, his stance on the environment is also cited as a factor in his political recovery.

7 In an interview with Marion Urban, Special Advisor to Heidemarie Wieczorek-Zeul, on 07-02-03 it was said that experts from the BMZ undertook dialogue with many others. It can safely be assumed that this would also been the case with most if not all of the German delegates.

8 Interview 21-02-03.

9 Interview 12-02-03.

10 Telephone interview with Wolfgang Sachs 27-05-03.

11 Information available at Europe Online states that there were 66 founding member and that as of 19 February 2003 78 countries had joined the coalition and more were expected to do so. www.europa.eu.int/comm/environment/climat/ johannesburg.htm May 2003. Other reports give higher numbers, for example 'some 100' is stated in Wille, Joachim 'World Summit: Strategies for the Future' *Deutschland* 5 (2002). www.publikation-deutschland.de/content/archiv-eng/02-05/art3.html November 2002.

12 Telephone interview 27-05-03.

13 Telephone interview 27-05-03.

14 Interview 11-03-03.

15 Interview 10-03-03.

16 Karsten Sach headed the German delegation during the first week of COP8, prior to Minister Trittin's arrival. Sach is officially noted as Deputy Head of Delegation.

17 An article 'EU Furious over Weak UN draft on Climate Change' in Corpwatch India on 28-10-02 reports that the EU were unhappy 'with attempts to link climate change with the issue of sustainable development, as advocated by several developing countries and by international activists at a two-day parallel Climate Justice Summit' Thomas Becker from Denmark is quoted in the article and is noted as being an EU representative, he is cited as saying "We think in the EU that climate change affects developing countries the most" and "We don't want to get into a competition between climate change and other issues or any quid pro quo." The EU is supposed to present 'one voice' at summits such as COP8. Trittin's speech and the comments by Becker could be seen as showing a split in this voice.

18 The internal politics of Germany is worth mentioning here. During interviews it was asserted that the BMU wanted a definite 40% reduction target, but was restrained by the BMWA who argued that Germany was reaching the Kyoto target and that other countries had to share the burden, hence, the compromise position of 40% on the understanding that the EU committed to 30% reductions.

19 For further information see unfccc.int/sessions/workshop/0206a6/documents/web0204.pdf April 2004.

[20] Interview 08-04-03.
[21] Telephone interview 24-04-03.
[22] Interview 17-04-03.
[23] Interview 08-04-03.

7. Actions outside of WSSD and COP8

[1] Interview 08-04-03.
[2] For further information see www.gefweb.org March 2004.
[3] For more information see http://gefweb.org/Replenishment/Summary_of_negotiations_-_ENGLISH_11-5.doc September 2003.
[4] Interview 10-03-03.
[5] For more information see www.gefweb.org/Projects/projects-Projects/PROGLIST.pdf January 2004.
[6] The Wuppertal Institute is a member of the National Advisory Committee, which is a consulting group for the German government. The institute also organised several side events at the conference.
[7] Interview 17-04-03.
[8] Interview 08-04-03.
[9] Interviews with Karsten Sach 08-04-03 & Simon Crabbe 12-09-03.
[10] Interview with Peter Fischer 18-02-03.
[11] It is interesting to note that some States of the USA, including California, have expressed an interest in joining the JREC. At the time of writing this is not possible as only sovereign States may sign. There is some discussion as to whether it may be possible for the JREC to have a second list of signatories for non sovereign States, this could also include corporations. Source: Interview with Representative of DG Environment 24-02-04.
[12] Interview with Peter Fischer 17-04-03.
[13] Interview with Peter Fischer 18-02-03.
[14] Joint Initiative.
[15] For more information see www.gtz.de/climate/english/bysector.htm February 2003.
[16] Interview 10-03-03.
[17] Interview 24-02-04.
[18] Clearly the general editorial remit of the ENB is known by the German government.
[19] According to Patrick Nunn of the University of the South Pacific, some 90% of Fiji's population live on the coastal flats, this is also where most economic activity occurs. www.rnw.nl/hotspots/html/tuv020828.html April 2004.
[20] UN Department of Public Information. *Small Islands Press Release August 1999.*
[21] For further information see Jaggard, Lyn. *The Influence of Small States: What means are available to low lying small island states of the Pacific to offset the threats of climat change?'* (University of Nottingham, 2001).
[22] For further information see www.usp.ac.fj/marine/msp_train2.htm April 2004.

[23] Tokelau was not included in the original proposal, which was approved on 1 February 2002, but is included in the project, implementation of which began in May 2003. For further information see: www.gefonline.org/projectDetails.cfm?projID=1058 May 2004 and South Pacific Environment Programme, 2003.

[24] South Pacific Environment Programme, 2003.

[25] Interview 27-05-04. M.J.Mace is Programme Director of the Climate Change and Energy Programme at the Foundation for International Environmental Law and Development; she provides advice and assistance to AOSIS with regard to the UNFCCC process. Previously M.J.Mace was Assistant Attorney General for the Federated States of Micronesia (FSM) Department of Justice; she has represented the FSM at numerous climate change negotiations. (www.field.org.uk May 2004).

[26] Interview 08-04-03.

[27] Personal correspondence with Karsten Sach May 2004.

[28] Interview 10-03-03.

[29] Interview 25-03-03.

[30] Personal correspondence April 04. Lefevere was present at the WSSD as a FIELD representative.

[31] For further information see www.usp.ac.fj/marine April 2004.

BIBLIOGRAPHY

Primary Sources

Documents and reports

BMU. *Declaration of like-minded countries on increasing the share of renewable energies.* September 2002 www.bmu.de/english/download/b_jointdeclaration.php October 2002

BMU. (2002) *German Environment Report 2002: Report on environmental policy in the 14th legislative period.* Berlin. BMU.

BMU. (2000) *Germany's National Climate Protection Programme. Summary.* Berlin. BMU.

BMU. *The Federal Environmental Agency.* www.bmu.de/english/tasks/uba.htm October 2001

BMZ. (2002a) *From Rio to Johannesburg: A review of progress in selected areas of German development policy.* Berlin. BMZ. www.bmz.de/include/cgibin/druck.pl?default August 2002.

BMZ. (2002b) *Establishment of priorities in development co-operation.* Bonn. BMZ.

Commission of the European Communities – Directorate General XI Environment, Nuclear Safety and Civil Protection. (1993) *Administrative Structures for Environmental Management in the European Community.* Luxembourg. Office for Official Publications of the European Communities.

Commission of the European Communities. (1984) *Ten Years of Community Environment Policy.* Brussels. Commission of the European Communities.

Eid, Uschi. (2001) 'Key Issues On The Road to Earth Summit 2002' *From Rio to Johannesburg: Contributions to the Globalization of Sustainability. World Summit 2002 Johannesburg Papers No. 5.* Berlin. Heinrich Böll Foundation.

European Commission. (2002a) *The EU agenda for the World Summit on Sustainable Development.* http://europa.eu.int/comm/environment/wssd/documents/agenda-en.pdf July 2002

European Commission. (2002b) *The World Summit on Sustainable Development: People, planet, prosperity*. Luxembourg. Office for Official Publications of the European Communities.

European Commission. (2001) *Environment 2010: Our Future, Our Choice: 6th EU Environment Action Programme 2001-2010*. Luxembourg. Office for Official Publications of the European Communities.

European Environment Agency. (2004) 'Energy subsidies and renewables.' *EEA Briefing No: 2/2004*. Copenhagen. European Environment Agency. http://reports.eea.eu.int/briefing_2004_2/en/Briefing_energy_EN_web.pdf June 2004

Federal Government. *"Prospects for Germany" – agreement on federal government's national sustainability strategy*. July 2002 http://eng.bundesregierung.de/documente/Artikel/ix_7632.htm November 2002

Federal Government. (2002) *Third Report by the Government of the Federal Republic of Germany in accordance with the Framework Convention of the United Nations*. Berlin. Federal Government.

Forum Umwelt und Entwicklung. (2002) *Globale Gerechtigkeit ökologisch gestalten: Neuer Schwung für nachhaltige Entwicklung*. Bonn. Forum Umwelt und Entwicklung.

Forum Umwelt und Enticklung (2002) *Zukunft für alle! Mitmachen-Bewegen-Verändern*. Bonn. Forum Umwelt und Entwicklung. www.rio-10.de/berliner_aktionstage.html August 2002

Germanwatch *The Climate Responsibility Campaign: Our Climate is drowning –. Who is Responsible?* www.germanwatch.org/klak/intro-e.htm November 2002.

Germanwatch *Klimakarawane*. www.germanwatch.org/klak/karawane.htm November 2002.

Global Environment Facility (2002) *Summary of negotiations on the third replenishment of the GEF Trust Fund*. http://gefweb.org/Replenishment/Summary_of_negotiations_-_ENGLISH_11-5.doc - September 2003.

Heinrich Böll Foundation. (2001) From Rio to Johannesburg: Contributions to the Globalization of Sustainability. World Summit 2002 Johannesburg Papers No. 5. Berlin. Heinrich Böll Foundation.

Marrakesh Ministerial Declaration *www.unfccc.int* November 2001

Nick-Leptin, Joachim. *Bundeskanzleramt letter to Geschäftsführer des Forums Umwelt und Entwicklung* – 13-08-02. www.rio-10.de/rioprozess/texte/kanzleramt_antwort_%20130802.pdf November 2002.

OECD Environment Directorate. (2002) *OECD Environmental Data Compendium 2002*. www.oecd.org/dataoecd/8/62/2958142.pdf October 2004

Ott, Hermann. *et al.* (2004) *South – North Dialogue on Equity in the Greenhouse: A proposal for an adequate and equitable global climate agreement.* Wuppertal. Wuppertal Institute for Climate Environment and Energy & University of Cape Town.

Trittin, Jürgen. (2001) 'Towards a Global Sustainable Energy Policy' in *From Rio to Johannesburg: Contributions to the Globalization of Sustainability. World Summit 2002 Johannesburg Papers No. 5.* Berlin. Heinrich Böll Foundation.

UN. (2002) *Report of the International Conference on Financing and Development. Monterrey, Mexico. 18-22 March 2002.* New York. UN.

UNCED. (1992) *The Global Partnership for Environment and Development: A Guide to Agenda 21.* Geneva. UNCED.

UNEP *UNEP's Executive Director Klaus Töpfer.* www.unep.org/OED/Topfer/default.asp May 2002

UNFCCC (2003) *Report of the Conference of the Parties on its eighth session, held at New Delhi from 23 October to 1 November 2002. Addendum. Part Two: Action taken by the Conference of the Parties at its eighth session.* http://unfccc.int/resource/docs/cop8/07a01.pdf November 2003

UNFCCC (2002a) *Financial Mechanism: Review of the Financial Mechanism: Compilation of submissions on the review of the financial mechanism.* http://unfccc.int/resource/docs/2002/sbi/misc02.pdf January 2004

UNFCCC (2002b) *Draft Decision -/CP.8: Review of the financial mechanism.* http://unfccc.int/cop8/latest/8_sbi.pdf January 2004

UNFCCC (1999) *Report on the in-depth review of the second national communication of Germany.* www.unfccc.de April 2002.

UNFCCC (1997) *Report on the in-depth review of the national communication of Germany.* www.unfccc.de April 2002.

UNFCCC. Climate Change Information Sheet 18 www.unfccc.de/resource/iuckit/fact18.html October 2001

WBGU (2001) *The Johannesburg Opportunity: Key Elements of a Negotiation Strategy.* Berlin. WBGU.

WBGU (1999) *Welt im Wandel: Umwelt und Ethik: Sondergutachten 1999.* Marburg. Metropolis-Verlag.

WSSD (2002) *The Johannesburg Declaration on Sustainable Development. 04-09-02.* www.johannesburgsummit.org/html/documents/summit_docs/1009wssd_pol_declaration.doc October 2002.

WSSD (2002) *World Summit on Sustainable Development Plan of Implementation.* www.johannesburgsummit.org/html/documents/summit_docs/2309_planfinal.htm April 2003

Wuppertal Institute for Climate, Environment and Energy. (2003a) *South-North Dialogue – Equity in the Greenhouse. Project Description.* Wuppertal.

Wuppertal Institute for Climate, Environment and Energy. (2003b) *Emissionshandel und Joint Implementation – der Stand der Dinge. Zusammenfassung des Workshops vom 13 – 14 März 2003 in Leipzig.* Wuppertal. www.wupperinst.org/download/TerraTec-2003/TerraTec_Final.pdf March 2003

Interviews
Bals, Christoph. Germanwatch. 11-03-03. Bonn.
Biermann, Frank. Potsdam Institute for Climate Impact Research. 25-03-03. Potsdam.
Blobel, Daniel. Ecologic. 12-02-03. Berlin.
Brockhagen, Dietrich. WBGU. 03-03-03. Berlin.
Budhooram, June. UNFCCC. 11-03-03. Bonn.
Crabbe, Simon. Department of Environment Food and Rural Affairs. 12-09-03. London.
Diaz-Bone, Harald. UNFCCC. 11-03-03. Bonn.
Fischer, Peter. Auswärtiges Amt. 18-02-03 and 17-04-03. Berlin.
Frisch, Thomas. BMWA. 24-04-03. By telephone.
Fuentes Hutfilter, Ursula. WBGU. 03-03-03. Berlin.
Görlach, Benjamin. Ecologic. 12-02-03. Berlin.
Hare, Bill. Greenpeace. 25-03-03. Potsdam.
Hein, Joachim. BDI. 22-04-03. Berlin.
Hey, Christian. Umweltrat. 13-02-03. Berlin.
Knill, Philipp. BMZ. 10-03-03. Bonn.
Kononov, Sergey. UNFCCC. 11-03-03. Bonn.
Krämer, Andreas. Ecologic. 12-02-03. Berlin.
Mace, M.J. FIELD. 27-05-04. London.
Matsarski, Vitaly. UNFCCC. 11-03-03. Bonn.
Mez, Lutz. Freie Universität. 03-02-03. Berlin.
Oberthür, Sebastian. Ecologic. 20-02-03. Berlin.
Perdomo, Martha. UNFCCC. 12-03-03. Bonn.
Representatives (2) of the Energy and Transport Directorate General of the European Commission. 24-02-04. Brussels.
Representative of the Environment Directorate General of the European Commission. 24-02-04. Brussels.
Sach, Karsten. Head of Division for International Co-operation, Global Conventions, and International Climate Change. BMU. 21-02-03 and 08-04-03. Berlin.
Sachs, Wolfgang. Wuppertal Institute for Climate, Environment, and Energy. 27-05-03. By telephone.
Sartorius, Rolf. Umweltbundesamt. 14-02-03. Berlin.
Siebenhüner, Bernd. Potsdam Institute for Climate Impact Research and Oldenburg Universität. 06-02-03. Berlin.

Simeonova, Katia. UNFCCC. 11-03-03. Bonn.
Suplie, Jessica. BMU. 06-02-03. Berlin
Treber, Manfred. Germanwatch. 12-03-03. Bonn.
Urban, Marion. Special Advisor to Heidemarie Wieczorek-Zeul, BMZ. 07-02-03. Berlin.
Weiss, Martin. Umweltbundesamt. 14-02-03. Berlin.
Wenderoth, Jürgen. BMF. 08-04-03. Berlin.

Personal correspondence
Knill, Philipp. German Embassy La Paz, ex of BMZ. November 2004.
Lafeld, Sascha. Dresdner Bank. April 2003.
Lefevere, Jürgen. Ex FIELD. April 2004.
Liptow, Holger. GTZ. March 2003.
Oberthür, Sebastian. Ecologic. November 2004.
Ott, Hermann. Wuppertal Institute for Climate Environment and Energy. May 2004.
Sach, Karsten. BMU. May 2004.
Seifried, Rolf. KfW. February 2003.
Suplie, Jessica. BMU. November 2004.
Vorrath, Judith. SEF. November 2004.

Press articles and media releases
BBC. *Germany Boosts Wind Power* http://news.bbc.co.uk/hi/english/world/europe/newsid_1765000/1765054.stm January 2002
BBC Monitoring Global Newsline European Political File. *German minister defends Johannesburg summit energy compromise* 03-09-02. Source quoted as ddp news agency, Berlin, in German 0711 gmt 03-09 02
BBC via Financial Times. *Chancellor says Germany to fight for more "Global Justice".* 13-05-02. Source quoted as globalarcive.ft.com/globalarchive/article.html?id=020513005822&query-World+Summit+ob+Sustainable+Development May 2002
Berliner Morgenpost. *Beifall für Schröder, Pfiffe für Powell.* 05-09-02 Page 6.
Berliner Morgenpost. *Klimaschutz: Trittin widerspricht den USA.* http://morgenpost.berlin1.de/inhalt/politik/story559167.html November 2002
BMU. *First German-Chinese environment forum opens in Beijing.* Press statement 239/03. Berlin 19 December 2003. www.bmu.de/en/1024/js/news/pressrelease031219/main.htm December 2003
BMU. *German participants arrange a CO_2-neutral trip to Johannesburg.* Press statement 204/02. Berlin 26 August 2002 www.bmu.de/english/news/pressrelease020826.php August 2002

BMU. (2001) 'Germany: helping save the environment of a surging economy' in *Common Ground.* 1/2001. Berlin. BMU.

BMU. *Jürgen Trittin: Renewable energies the key for a future-oriented global energy supply.* Press statement 216/02. Johannesburg/Berlin 3 September 2002 www.bmu.de/english/news/pressrelease020903.php October 2002

BMU. *Jürgen Trittin: Solar power stations – Germany's export hit of the future.* at www.bmu.de/english/news/pressrelease020314.php March 2002

BMU. *State of the World 2002* www.bmu.de/english/news/pressrelease020115.php January 2002

Bonner General Anzeiger. Merey, C. *Der Sieg der "unheiligen Allianz"* 04-09-02. Page 4.

Bonner General Anzeiger. *Viel Lob für Idee einer Energiekonferenz in Bonn* 05-09-02. Page 5.

Climate Action Network. *Eco* 01-11-02 Vol. CVIII:10

Climate Action Network. *Eco* 29-10-02 Vol. CVIII:6

Climate Action Network. *Eco* 23-10-02 Vol. CVIII:1

Corpwatch India. *EU Furious over Weak UN draft on Climate Change.* 28-10-02 www.corpwatchindia.org/news/PRT.jsp?.articleid=3025 November 2002

Deutsche Welle. *Climate Conference Opens in India.* 23-10-02 http://dw-world.de/dwelle/cda/details/dwelle.cda..../ 0,3820,1446_AD_660413_A,00.htm October 2002

Der Spiegel Online. Kulick, H. *Energie-Krise beim Gipfel* www.spiegel.de/politik/ausland/0,1518,212161,00.html September 2002

Der Spiegel Online. Kulick, H. *Die Erdölländer spielen die Schwachen gegeneinander aus* www.spiegel.de/politik/ausland/0,1518,212084,00.html September 2002

Der Spiegel Online. *Verbrechen gegen die Menscheit* www.spiegel.de/politik/ausland/0,1518,212346,00.html September 2002

Der Tagespiegel. *Als hätte es das Kyoto-Protokoll nie gegeben; bei der Klimaschutzkonferenz verhindern USA neue Ziele.* 01.11.02 http://archiv.tagespiegel.de/archiv/01.11.2002/284545.asp November 2002

Der Tagespiegel. *Weltgipfel: Nein zur Energiewende.* 04-09-02 Page 1.

Die Tageszeitung. *Gipfel der Kompromisse.* 04-09-02 Page 1

Environment Daily 1219, *German green energy rules win EU clearance.* 22/05/02. www.climnet.org/news/May2002.htm May 2002

Environment Daily 1186, *German Parliament finally agrees to lorry charging.* 26/03/02. www.climnet.org/news/March2002.htm March 2002

Frankfurter Allgemeine Zeitung. *Deutschland scheitert auf dem Weltgipfel mit Vorschlägen zur Energiepolitik.* 04-09-02 Page 1.

Frankfurter Rundschau. *Umweltschützer loben den Kanzler* 03-09-02 Page 2.

German Press and Information Office. Press Release No. 588/01. *Minister of State Bury presents draft of national sustainability strategy containing 21 objectives for the 21st century.* 19-12-01.
http://eng.bundesregierung.de/dokumente/Pressemitteilung/ix_65804_2726.htm November 2002.

Independent News. *Earth Summit: After days of intense negotiations, leaders settle on a blueprint to keep the planet alive* 03-09-02
http://news.independent.co.uk/world/.../story.jsp?dir=507&story=329998&host=3&printable= September 2002

International Institute for Sustainable Development. *Earth Negotiations Bulletin COP-8 Final* 01-11-02, Vol. 12:209

International Institute for Sustainable Development. *Earth Negotiations Bulletin COP-8. Issue 8.* 01-11-02, Vol. 12:208

International Institute for Sustainable Development. *Earth Negotiations Bulletin COP-8. Issue 7.* 31-10-02, Vol. 12:207

International Institute for Sustainable Development. *Earth Negotiations Bulletin COP-8. Issue 2.* 24-10-02, Vol. 12:202

International Institute for Sustainable Development. *Earth Negotiations Bulletin. WSSD – Issue 4.* 29-08-02 Vol.22:45

International Institute for Sustainable Development. *Earth Negotiations Bulletin on the* side. *WSSD - Issue 2.* 28-08-02.
www.iisd.ca/linkages/2002/wssd/enbots August 2002.

International Institute for Sustainable Development. *Earth Negotiations Bulletin on the side, WSSD - Issue 6.* 02-09-02.
www.iisd.ca/linkages/2002/wssd/enbots September 2002.

International Institute for Sustainable Development. *Earth Negotiations Bulletin on the side, WSSD -Issue 8.* 03-09-02.
www.iisd.ca/linkages/2002/wssd/enbots September 2002.

International Institute for Sustainable Development. *Earth Negotiations Bulletin, WSSD PrepCom IV – Issue 10.* Vol.22:39. 06-06-02.

International Institute for Sustainable Development. *Earth Negotiations Bulletin, WSSD PrepCom IV – Issue 1.* Vol.22:30. 27-05-02.

Johannesburg World Summit Company. *Media Release. German Government funds Summit Preparations.* 01 Aug 02. from Jowsco. Email@mslpr.za August 2002

Online.ie *Schroeder pledges E1bn to renewable energy* 02-09-02
www.online.ie/news/viewer.adp?article=1823854 September 2002

Pacific Islands Forum Secretariat. *Press Statement 7702: Smaller Island states statement on climate change.* 15-08-02.
www.forumsec.rog.fj/news/2002/Aug10.htm August 2002

Planet Ark. *German Parliament ups subsidies for green energy.*
www.planetark.org/dailynewsstory.cfm/newsid/13358/story.htm November 2001

Planet Ark. *OPEC said blocking Summit "green" energy goal.* www.planetark.org/avantgo/dailynewsstory.cfm?newsid=17559 September 2002

Planet Ark. *Tuvalu seeks help in US global warming lawsuit.* 30-08-02 www.planetark.org/avantgo/dailynewsstory.cfm?newsid=17514 September 2002

Trittin, Jürgen. (2002) 'Johannesburg Summit produces a positive overall result' in *Common Ground.* 3/2002. Berlin. BMU. www.bmu.de/infos/magazine/common/com021100/aword.php October 2003

UN Department of Public Information. *Small Islands Press Release August 1999.*

WBGU (1999) *Ethics is indivisible: A plea for responsible treatment of nature.* Press release – 17-12-99. Berlin. WBGU. www.wbgu.de/WBGU/wbgu_sn1999_presse_engl.html November 2002

Wille, Joachim. 'World Summit: Strategies for the Future.' *Deutschland.* Issue 5/2002. www.publikation-deutschland.de/content/archiv-eng/02-05/art3.html November 2002

Yahoo News. Doyle, Alister. & Ginsberg, Jodie. *Earth Summit Agrees on Energy, Angers Greens.* 02-09-02 http://story.news.yahoo.com/news?tmpl=story&u=/nm/.../canada_environment_summit_col_ September 2002

Speeches

Schröder, Gehard. *Statement by Mr. Gerhard Schröder, Chancellor of the Federal Republic of Germany at the World Summit on Sustainable Development* 02-09-02 www.un.org/events/wssd/statements/germanyE.htm September 2002

Trittin, Jürgen. *The Success Story of Climate Protection in Germany.* 01-11-02 www.bmu.de/english/topics/speeches/speech_trittin_021101.php November 2002

Trittin, Jürgen. *Climate Change and Sustainable Development – two sides of the same coin.* Statement at the Ministerial Round Table, Session II. 31-10-02 www.bmu.de/english/topics/speeches/speech_trittin_021031.php November 2002

Trittin, Jürgen. *Sustainable Development and the Role of Solar Thermal Power.* 03-09-02 www.bmu.de/reden/rede_trittin020903.php October 2002

Volmer, Ludger. *Strengthening the OSCE's Role in the Realm of Environment and Security.* 03-07-01. www.auswaertiges-amt.de/www/en/ausgabe_archi_id=1767 October 2004

Wolf, Margareta. *Eröffnungstatement.* Workshop zu Industriepolitiken 2. und 3. Dezember 2002. Berlin. 02 12 02 www.bmu.de/reden/rede_wolf021202.php April 2003

URLs
www.bmz.de/en/topics/umwelt/umwelt5.pdf February 2004
www.bundesregierung.de/en/News-by-subject/International-,11075/Summits-and-Conferences.htm August 2002
www.ciesin.org November 2001
europa.eu.int/abc/treaties_en.htm November 2004
europa.eu.int/comm/environment/climat/gge_press.htm July 2003
www.field.org.uk May 2004
www.forumue.de/aboutus/secretariat/index.html October 2002
www.gefonline.org/projectDetails.cfm?projID=1058 May 2004
www.gefweb.org March 2004
www.gefweb.org/Outreach/outreach-PUblications/Project_factsheet/Asia_Pacific-paci-3-cc-undp-eng.pdf April 2004
www.gefweb.org/Projects/projects-Projects/PROGLIST.pdf January 2004
www.gtz.de/climate/english/bysector.htm February 2003
www.mep-online.com/chapter2/section_2_9_10.html September 2004
www.nachhaltigkeitsrat.de/service/download_e/pdf/Perspectives_for_Germany_short_version_pdf December 2002
www.renewables2004.de/en/2004/particpants.asp?page=print October 2003
www.rio-10.de/berliner_aktionstage.html August 2002
www.rio-10.de/presse/pm_offener_brief060802.html August 2002
www.rio-10.de/rioprozess/texte/10_points_for_johannesburg.html August 2002
www.rio-10.de/rioprozess/texte/kanzleramt_antwort_%20130802.pdf August 2002
www.rnw.nl/hotspots/html/tuv020828.html April 2004
www.sef-bonn.org.en/about/profile/index.php?print=3 October 2004
www.sidsnet.org/aosis/background.htm November 2001
www.sidsnet.org/aosis/main-aosis.htm November 2001
www.un.org/esa/sustdev/csd/ecn172001.19e.htm February 2004
www.un.org/Overview/rights.html June 2002
www.unece.org/env/pp/welcome.html November 2002
www.unfccc.de/reso...process/components/response/respkp.html November 2001
www.usp.ac.fj/marine/ April 2004
www.usp.ac.fj/marine/msp_train2.htm April 2004

Secondary Sources

Books

Anderson, Michael. S. & Liefferink, Duncan. (1997) 'Introduction: the impact of the pioneers on EU environmental policy.' In Anderson, M.S. & Liefferink, D. (eds) *European Environmental Policy: The Pioneers.* Manchester. Manchester University Press.

Beuermann, Christiane. & Burdick, Bernhard. (1997) 'The Sustainable Transition in Germany: Some Early Stage Experiences.' in O'Riordan, Timothy. & Voisey, Heather. (eds) *Sustainable Development in Western Europe: Coming to Terms with Agenda 21.* London. Frank Cass & Co Ltd.

Beuermann, Christiane. & Jäger, Jill. (1996) 'Climate Change Politics in Germany: How long will any double dividend last?' in O'Riordan, Timothy & Jäger, Jill. (eds) *Politics of Climate Change: A European Perspective.* London. Routledge.

Bichsel, Anne. (1996) 'NGOs as agents of public accountability and democratization in intergovernmental forums.' in Lafferty, William. M. & Meadowcroft, James. (eds) *Democracy and the Environment.* Cheltenham. Edward Elgar Publishing Ltd.

Bretherton, Charlotte. & Vogler, John. (1999) *The European Union as a Global Actor.* London. Routledge.

Clark, Ian. (1989) *The Hierarchy of States: Reforms and Resistance in the international order.* Cambridge. Cambridge University Press.

Doyle, Timothy & McEachern, Doug. (1998) *Environment and Politics.* London. Routledge.

Economy, Elizabeth. & Schreurs, Miranda. A. (1997) 'Domestic and international linkages in environmental politics.' in Schreurs, Miranda. A. & Economy, Elizabeth. (eds) *The Internationalization of Environmental Protection.* Cambridge. Cambridge University Press.

Green, Simon. & Paterson, William. E. (2005) 'Introduction: Semisovereignty challenged.' in Green, Simon. & Paterson, William. E. (eds) *Governance in Modern Germany.* Cambridge. Cambridge University Press.

Grix, Jonathan, (2004), *Foundations of Research.* London. Palgrave Macmillan.

Habermas, Jürgen. (2002) 'The Tasks of a Critical Social Theory.' in Calhoun, Craig. Gerteis, Joseph. Moody, James. Pfaff, Steven. & Virk, Indermohan. (eds) *Contemporary Sociological Theory.* Malden, MA. Blackwell Publishing Ltd.

Habermas, Jürgen. (translated by Rehg, William.) (1996) *Between Facts and Norms: Contributions to a Discourse Theory of Law and Democracy.* Cambridge. Polity Press.

Haigh, Nigel. (1996) 'Climate Change Policies and Politics in the European Community.' in O'Riordan, Timothy. & Jäger, Jill. *Politics of Climate Change: a European Perspective.* London. Routledge.

Harber, Katherine. & Payton, Geoffrey. (eds) (1989) *Heinemann English Dictionary: New Edition.* Oxford. Heinemann Educational.

How, Alan. (2003) *Critical Theory.* Basingstoke & New York. Palgrave Macmillan.

Hughes, John. & Sharrock, Wesley. (1997) *The Philosophy of Social Research. Third Edition.* London & New York. Longman Limited.

Katzenstein, Peter.J. (1987) *Policy and Politics in West Germany: The Growth of a Semisovereign State.* Philadelphia. Temple University Press.

Kegley, Charles. W. Jr & Wittkopf Eugene. R. (1999) *World Politics: Trend and Transformation. Seventh Edition.* Boston & New York. Worth Publishers Inc.

Keohane, Robert. O. & Nye, Joseph. (1989) *Power and Interdependence: World Politics in Transition. 2nd edition.* Boston. Little Brown & Co.

Lafferty, William. M. & Meadowcroft, James. (1996) *Democracy and the Environment.* Cheltenham. Edward Elgar Publishing Ltd.

Lees, Charles. (2005) 'Environmental Policy: the Law of Diminishing Returns?' in Green, S. & Paterson, W. E. (eds) *Governance in Modern Germany.* Cambridge. Cambridge University Press.

Lukes, Steven. (1974) *Power: A radical view.* London. Macmillan Press Ltd.

Maddock, Rowland. (1998) 'Environmental politics & policies in the Asia Pacific.' in Maidment, Richard. Goldblatt, David. & Mitchell, Jeremy. (eds) *Governance in the Asia Pacific.* London. Routledge.

(Rosenau referenced as Rosenau J. 'Governance in the twenty-first century' *Global Governance.* 1995 Vol.1. No.1.)

Mason, Michael. (1999) *Environmental Democracy.* London. Earthscan Publications Ltd.

Mayer, Margil. & Ely, John. (eds) (1998) *The German Greens: Paradox between movement and party.* Philadelphia. Temple University Press.

McCarthy, Thomas. (1984) *The Critical Theory of Jürgen Habermas.* Cambridge. Polity Press.

Mewes, Horst. (1998) 'A Brief History of the German Green Party.' in Mayer, Margil. & Ely, John. (eds) *The German Greens: Paradox between movement and party.* Philadelphia. Temple University Press.

Paterson, Matthew. (2001) 'Principles of Justice in the Context of Global Climate Negotiations.' in Luterbacher, Urs. & Sprinz, Detlef. (eds) *International Relations and Global Climate Change.* Cambridge, MA. The MIT Press.

Paterson, Matthew. (1996) *Global Warming and Global Politics.* London. Routledge.

Rossi, Philip. (1998) 'Public Argument and Social Responsibility: The Moral Dimension of Citizenship in Kant's Commonwealth.' in Kneller, Jane. & Axinn, Sydney (eds) *Autonomy and Community; Readings in*

Contemporary Kantian Social Philosophy. Albany. State University of New York Press.
Rowlands, Ian. (1995) *The politics of global atmospheric change.* Manchester. Manchester University Press.
Schelling, Thomas. C. (1960) *The Strategy of Conflict.* Cambridge, MA. Harvard University.
Schreurs, Miranda. A. (1997) 'Domestic institutions and international environmental agendas in Japan and Germany.' in Schreurs, Miranda. A. & Economy, Elizabeth. (eds) *The Internationalization of Environmental Protection.* Cambridge. Cambridge University Press.
Shue, Henry. (1980) *Basic Rights: Subsistence, Affluence and US Foreign Policy. 2nd Edition.* Princeton. Princeton University Press.
Stokke, Olav. S. (1998) 'Understanding the Formation of International Environmental Regimes: The Discursive Challenge.' in Underdal, Arild. (ed) *The Politics of International Environmental Management.* Dordrecht. Kluwer Academic Publishers.
Sturm, Roland. (1996) 'Continuity and Change in the Policy-Making Process.' in Smith, Gordon. Paterson, William. & Padgett, Stephen. (eds) *Developments in German Politics 2.* Basingstoke. Macmillan Press Ltd.
Von Weizsäcker, Ernst. (1994) *Earth Politics.* London. Zed Books Ltd.
Weale, Albert. (1992) *The new politics of pollution.* Manchester. Manchester University Press.
Wurzel. Rüdiger. (2002) *Environmental policy-making in Britain, Germany and the European Union: The Europeanisation of air and water pollution control.* Manchester. Manchester University Press.

Journal Articles
Brulle, Robert. J. 'Habermas and Green Political Thought: Two Roads Converging' *Environmental Politics.* Winter 2002. Vol.11:4, pp1-20
Cavalier, Robert. Mellon, Carnegie. & Ess, Charles. (2002) *Introduction to Habermas's Discourse Ethics.*
http://caae.phil.cmu.edu/cavalier/Forum/meta/background/HaberIntro.html
quoting Habermas, Jürgen. 'Reconciliation through the Public Use of Reason: Remarks on John Rawls's Political Liberalism' *Journal of Philosophy* March 1995. Vol.XCII:3, pp117-118
Cavender, Jeannine. & Jäger, Jill. 'The History of Germany's Response to Climate Change.' *International Environmental Affairs.* Winter 1993. Vol.5:1, pp.3-18.
Clayton, Susan. 'Models of Justice in the Environmental Debate.' *Journal of Social Issues.* 2000 Vol.56:3, pp459-474.
Hampson, Fen. O. 'Climate Change: building international coalitions of the like-minded.' *International Journal.* Winter 1989-1990 Vol.45, pp36-74.

Hatch. Michael. T. 'The Politics of Global Warming in Germany.' *Environmental Politics.* Autumn 1995. Vol. 4:3, pp.415-440.

O'Hara, Sabine. 'Discursive ethics in ecosystems valuation and environmental policy.' *Ecological Economics.* 1996. Vol.16, pp95-107.

Vogler, John. 'The European Union as an Actor in International Environmental Politics.' *Environmental Politics.* Autumn 1999. Vol.8:3, pp24-48.

Willetts, Peter. 'From Stockholm to Rio and beyond: the impact of the environmental movement on the United Nations consultative arrangements of NGOs. *Review of International Studies.* 1996 Vol.22, pp57-80.

Policy Papers and Reports

Biermann, Frank & Simonis, Udo. E. (1998) *A World Environment and Development Organization: Functions, Opportunities, Issues. Policy Paper 9.* Bonn. Development and Peace Foundation.

Doran, Peter. (2002) *World Summit on Sustainable Development (Johannesburg) – An assessment for IISD.* www.iisd.org/pdf/2002/wssd_assessment.pdf November 2002

Fues, Thomas. (1997) *Rio Plus 10: The German Contribution to a Global Strategy for Sustainable Development. Policy Paper 6.* Bonn. Development and Peace Foundation.

Guist, Christine (2002) *Germany and the World Summit on Sustainable Development in Johannesburg August 26 - September 4, 2002.* Bonn. Goethe Insitut.

Halpern, Shanna. 1992 UNCED: Process and documentation. Providence, RI: Academic Council for the United Nations System (ACUNS). Chapter One Preparations for UNCED p7, reproduced with permission www.ciesin.org/docs/008-585/unced-ch1.htm December 2001

Liptow, Holger. - GTZ (2002) *The Role of the CDM in German Climate Policy.* Presentation at Conference on EU and German Climate Policy – Hamburg May 2002. www.hwwa.de June 2002

Messner, Dirk & Nuscheler, Franz. (1996) *Global Governance: Challenges to German Politics on the Threshold of the Twenty-First Century. Policy Paper 2.* Bonn. Development and Peace Foundation.

Ott, Hermann. (2003) *Warning Signs from Delhi: Troubled Waters Ahead for Global Climate Policy.* www.wupperinst.org/download/Warning-Signs-Ott.pdf March 2003.

Ott, Hermann. & Takeuchi, Tsuneo. (eds) (2000) *Japan and Germany: Towards Co-ordinated Climate Protection Strategies.* Wuppertal. Wuppertal Institute for Climate, Environment and Energy.

Von Seht, Hauke. & Ott, Hermann. (2000) *EU environmental principles: Implementation in Germany*. Wuppertal. Wuppertal Institute for Climate, Environment and Energy.

Unpublished Sources

Jaggard, Lyn. (2001) *The Influence of Small States: What means are available to low lying small island states of the Pacific to offset the threats of climate change?* Thesis for Master of Research in Politics, University of Nottingham.

Lees, Charles. (2002) *Environmental policy: the Law of Diminishing Returns?* Paper presented at the University of Birmingham workshop 'Semisovereignty Revisited: Governance, Institutions and Policies in United Germany.' 26-28th April 2002.

Ulbert, Cornelia. (2003) *Ideas and Discursive Structures: U.S. and German Positions in Negotiating Framework Convention on Climate Change.* Paper presented at the 2003 Hamburg Conference 'Does Discourse Matter? Power, Discourse and Institutions in the Sustainability Transition.' 11-13th. July 2003.

INDEX